Topical
Antimicrobial
Testing
and
Evaluation

Topical Antimicrobial Testing and Evaluation

Daryl S. Paulson
BioScience Laboratories, Inc.
Bozeman, Montana

CRC Press
Taylor & Francis Group
Boca Raton London New York

CRC Press is an imprint of the
Taylor & Francis Group, an **informa** business

CRC Press
Taylor & Francis Group
6000 Broken Sound Parkway NW, Suite 300
Boca Raton, FL 33487-2742

First issued in paperback 2019

ISBN-13: 978-0-8247-1957-9 (hbk)
ISBN-13: 978-0-367-39979-5 (pbk)

Library of Congress Cataloging-in-Publication Data

Paulson, Daryl S.
    Topical antimicrobial testing and evaluation / Daryl Paulson.
      p.   cm.
    ISBN 0-8247-1957-3 (alk. paper)
      1. Antiseptics—Testing.   2. Antiseptics—Testing—Statistical methods.   I.   Title.
    RM400.P38   1999
    615′.7—dc21   99-14707
                                                                                           CIP

Visit the Taylor & Francis Web site at
http://www.taylorandfrancis.com

and the CRC Press Web site at
http://www.crcpress.com

Foreword

As Daryl Paulson says in the first chapter of this important book, "Because the goal is to introduce products that will be successful, manufacturers must develop a product from a multidimensional perspective." Simple as that sounds, it is exactly the failure to do so that has hampered both the development and the marketing of many worthwhile products. The reason is fairly simple: if reality is actually multidimensional, then failing to take that fact into account in any endeavor will cripple that endeavor. This is true from grand agendas, such as education and politics, to modest topical ones, including antimicrobials.

In a series of books (such as *A Brief History of Everything*), I have tried to develop the idea that any phenomenon can be looked at from both the inside and the outside, in both individual and collective forms. This gives us four major perspectives or dimensions: the interior of an individual (subjective, intentional), the exterior of an individual (objective, behavioral), the inside of the collective (intersubjective, cultural), and the exterior of the collective (interobjective, social systems).

For example, your consciousness can be looked at from the inside—the subjective side, your own awareness at this moment—which is experienced in the first person as an "I" (all the images, impulses, concepts, and desires floating through your mind right now). You can also study consciousness in an objective, empirical, scientific fashion, in the third person as an "it" (for example, the brain contains acetylcholine, dopamine, serotonin, and so forth, all described in objective it-language). And both of those exist not just in singular but in plural forms, not just as an "I" or an "it," but a "we." This collective form also has an inside and outside: the cultural values shared from within (e.g., morals, worldviews, cultural meaning) and the exterior concrete social forms seen from without (e.g., modes of production, technology, economic base, social institutions, information systems).

Since all four of those dimensions are very real and very significant factors, a failure to take them into account in product development is a severe handicap. The value of Daryl Paulson's book is that it shows precisely how and why a truly multidimensional approach can be brought to the development and marketing of antimicrobials. Of course, the same principles apply to business in general and indeed, to human endeavors altogether. But even on a small scale— and all endeavors start small—a genuinely multidimensional approach promises novel success.

Ken Wilber
author, philosopher

Preface

Over the years, the medical field has come to rely on topical antimicrobials to reduce the risk of infection. Currently, there are a variety of topical antimicrobials to choose from, as well as new application systems. It is important, therefore, that the measurement of antimicrobial effectiveness be accurate and precise, as well as valid. This book is an attempt to provide the rationale and the statistical designs most appropriate for measuring antimicrobial efficacy.

The book begins with an examination of the skin environment on which the microorganisms of importance reside, permanently or transiently. The most common topical antimicrobial formulations are then discussed, including chlorhexidine gluconate, providone iodine, parachlorometaxylenol (PCMX), triclosan, and alcohols. Next, basic experimental designs appropriate to evaluations of topical antimicrobial properties are described. These include surgical scrub formulation evaluations, healthcare personnel handwash evaluations, and preveinous/arterial catheter insertion preparations. Additionally, designs utilized to evaluate both consumer antimicrobial products and foodhandler handwash products are presented.

The last section of the book includes a discussion of basic statistical designs used in evaluating data from the testing of antimicrobial products. Both parametric and nonparametric approaches are described.

It is my hope that this book brings to both academia and industry the latest information in the field of topical antimicrobial efficacy evaluation.

ACKNOWLEDGMENTS

I have been especially influenced by the philosophical writings of Ken Wilber, the leading intellectual influence in integral modeling, which I adopted in taking a multiple perspective approach to developing topcial antimicrobial products. I am deeply indebted to my co-workers at BioScience Laboratories, Inc., to keep things running, while I wrote and rewrote the chapters in this book. They include

Carol Riccardi, Executive Director of Operations; Chris Beausoleil, Manager of Clinical Trials; Terri Eastman, Manager of In Vitro Laboratories; Michael Douglas, Director of Sales and Marketing; and my very capable and supportive wife, Marsha Paulson, Vice-President of BioScience Laboratories, Inc.

I am especially indebted to John Mitchell, Director of Quality Assurance, for his many hours of challenging my assumptions, utilizing his vast knowledge of microbiology, and editing this work. Tammy Anderson provided valuable assistance in the development of this book by typing and retyping it, making format changes, editing, creating figures and diagrams and, basically, managing the entire book development process. Her abilities are astounding.

I also thank the staff at Marcel Dekker, Inc.—particularly Brian Black and Maria Allegra—for its flexibility, professionalism, and quality concerns.

I want to thank my parents, Orrin and Sylvia, for standing behind me, especially during my dark years immediately following my return from combat in South Vietnam. I wish Orrin were alive to see this work, and I thank him for telling me that I could do whatever I wanted, as long as I worked hard enough to do it. He was right.

Finally, I dedicate this book to my fellow Marines who fought and died fighting in Vietnam . . . *Semper Fidelis.*

Daryl S. Paulson

Contents

1

General Approach to Developing Topical Antimicrobials

The intense levels of competition present in the topical antimicrobial market, as well as the ever-tightening Food and Drug Administration's (FDA) product performance standards demand that antimicrobial soap and detergent manufacturers produce products which meet market requirements.[1] But just what are the market requirements? They include such factors as antimicrobial effectiveness, low skin irritation, ease of use, aesthetics, and various other attributes. If these factors have been addressed, the product has probably been developed with concern and competence. But too often, manufacturers ignore these important factors and get a product to market merely to compete with those of other competitors. In the end, this approach often fails; the product is never really accepted into the market.[2] Because the goal is to introduce products that will be successful, manufacturers should develop a product from a multidimensional perspective.

No one has argued the importance of viewing the human situation from a multidimensional perspective more convincingly than human science theorist Ken Wilber.[3,4,5] Wilber, the architect of the holonic quadrant model, states that at least four perspectives should be addressed: social, cultural, personal objective, and personal subjective.

I. THE HOLONIC QUADRANT MODEL

A. Social Requirements

Social requirements, as they apply here, include conforming to the standards of regulating agencies such as the FDA, the Federal Trade Commission (FTC), and the Environmental Protection Agency (EPA), as well as the rules, laws, and regulations they enforce. Before completing the design of a product, it is critical to understand the regulations governing the product, the product's components

and their concentrations, as well as product stability and toxicological properties. For example, a New Drug Application (NDA) is required in order to market
a regulated drug product. For over-the-counter (OTC) products, the active drug
must be allowable and its levels within allowable limits. Additionally, the
FDA's Tentative Final Monograph for OTC products or the "to-be-determined"
requirements of the Cosmetic Toiletries and Fragrance Association (CTFA) and
the Soap and Detergent Association (SDA) must be addressed.

B. Cultural Requirements

Cultural requirements are very important, but are often ignored. Cultural and
subcultural requirements include shared values, beliefs, goals, and the world
views of a society or subgroup of society.[6] Shared values such as perceived
antimicrobial "effectiveness" have a large influence on consumers.[7] These values are generally of two types: manifest and latent.[3,5] Manifest (surface) values
are conscious to the consumer. For example, a consumer buys an antimicrobial
soap to be "cleaner" than they can be using a non-antimicrobial soap. But deeper
and more fundamental values lie behind those that are manifest. These are referred to as latent values and are unconscious to consumers in that they are
generally unaware of them. In this case, "cleaner" may encompass such basic
needs as being accepted, loved, and worthwhile as a person, spouse, and/or
parent.

  Most manifest and latent values we share as a culture are magnified by
manufacturers' advertising campaigns where consumers are frequently motivated by both manifest and latent values. For example, if a homemaker per-
ceives that by using antimicrobial soaps the children are being better taken care
of (a manifest value), and if the family makes the homemaker feel more valued,
more loveable, and/or more needed (latent values), the person will be motivated
to purchase the product. Finally, much of what consumers believe to be true is
not grounded in objective reality.[8] Most of these beliefs are formed from their
interpretation of mass media reports, opinions of others, and explanations of
phenomena from various notorieties.[10,11]

C. Personal Objective Attributes

Physical components of a product include its application, its characteristics, its
antimicrobial actions, its irritation effects to skin, and its in-use effects (e.g.,
staining clothing, gowns, and bedding). It is important that products be designed
with the individual in mind.[2] Hence, products must be easy to open (if in a
container), to use, and must be effective in its intended use (e.g., by food-
servers, home consumers, medical personnel, and surgical personnel).

D. Personal Subjective Attributes

This category includes one's personal interpretation of cultural and subcultural "world views." Relative to antimicrobials, these include subjective likes and dislikes of characteristics such as the fragrance and feel of the product, the perceived "quality" of the product, and other aesthetic considerations.[7] As with cultural attributes, manifest and latent values are operative in this category. Hence, if one likes the springtime fragrance of a consumer bodywash product (manifest), the latent or deeper value may be that it makes one feel younger and, therefore, more physically attractive and desirable as a person.

These four attribute categories can be presented in quadrant form (Figure 1).[3,4,5] Each quadrant interacts with the other three quadrants. For example, cul-

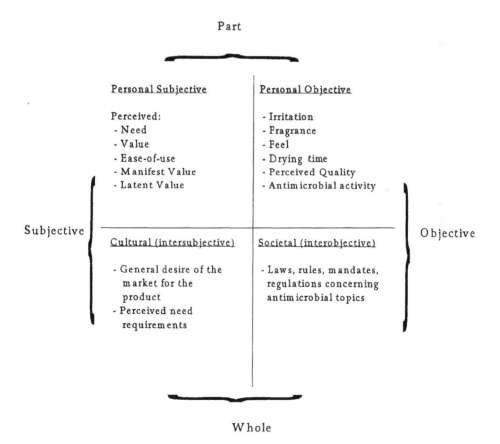

Figure 1 Quadrant model of attribute categories.

tural values influence personal values, and vice versa. Cultural and personal values influence behavior, and behavior influences values.*

II. A CHRONIC PROBLEM: REDUCTIONISM

A chronic problem of product design, reductionism occurs when personnel such as engineers, chemists, microbiologists, statisticians, and lawyers focus solely on the objective parameters (the two right quadrants) and, therefore, interpret subjective parameters in objective terms. That is, they reduce the subjective domains to objective ones.[3,4,5] Equally problematic are marketing personnel, industrial psychologists, and social scientists who focus their efforts in the subjective domains and interpret objective domains in subjective terms. Reductionism is often rooted in subcultural bias. Those who were trained in the physical sciences tend to have incorporated an objective "world view," while those trained in social sciences usually have a subjective one. Both world views are true, but only partially so, and both must be recognized and integrated into the product's development. This in no way suggests that each specialist work in all four quadrant domains; that is unrealistic. Each should focus in their own domain (e.g., chemists in two right quadrants, and marketing in the two left quadrants). However, the person or group in charge of the product's development must be conscious of both domains and integrate them into the design of the product. With this in mind, let us examine how the quadrant model may be of assistance in the design of topical antimicrobial products.

A. Product Categories

There are three general categories of topical antimicrobial products recognized by the FDA:

1. Preoperative skin-prepping formulations
2. Surgical scrub formulations
3. Healthcare personnel handwash products

The CTFA/SDA groups have recently introduced what they term the "Healthcare Continuum Model," adding three product categories to these:

1. Antimicrobial bodywashes
2. General-use antimicrobial handsoaps
3. Foodhandler handsoaps

---

*For a very complete discussion of this topic, readers are urged to consult the work of theorist Ken Wilber.

Whether these three additional categories will be formally recognized by the FDA remains to be seen, but because they are anchored in personal and cultural values, they are important. It is fair to say that most Americans want to feel clean and be clean. This, to many, means using antimicrobial bodysoap for showering and bathing, as well as for handwashing.

Moreover, there is considerable pressure on government agencies to make the food we eat safer. Consumers remember clearly the deadly problems created by ingesting *Escherichia coli*, Serotype O157/H7, in contaminated hamburger.[12] They want governmental protection to prevent this from recurring.[13,14]

B. Health Care Continuum Model

Let us now focus on the categories addressed in the Healthcare Continuum Model. As previously stated, the FDA acknowledges the healthcare personnel handwash, the surgical scrub, and the preoperative skin preparation evaluations. It is likely they will designate a separate "foodhandler handwash" category, but will probably not, at least in the near future, officially recognize the general-use antimicrobial handsoaps and antimicrobial bodywashes. With this said, let us turn our attention to development of products specific to each category.

C. Preoperative Skin-Prepping Formulations

1. Personal Objective (Upper Right Quadrant)

Preoperative skin preparation formulations are designed to degerm an intended surgical site rapidly, as well as provide a high level of persistent antimicrobial activity—up to six hours—post-skin prepping.[16] In terms of the quadrant model presented earlier (Figure 1), these requirements belong to the personal objective region (upper right quadrant).

The preoperative skin preparation product should be convenient and easy to use for the surgical staff consumers. Given two products with equal antimicrobial efficacy, one that is easier to use (e.g., has a shorter skin-prep time, can be more easily seen on the skin due to a coloring agent, and/or is a one-step procedure) will be preferred over the other product.

An often overlooked aspect of product acceptance is customer service. It is ironic that many firms spend large sums of money to assure a product meets its intended purpose, but fail to meet a customer's needs. Orders are delayed or lost, return phone calls are promised, but not made, and so on. In short, customers are ignored. Tom Peters points out that customers who find it hard to do business with manufacturers and vendors do not remain customers, even if the products provided are "better" than those of competitors.[16] Service is critically important.

2. Social Requirements (Lower Right Quadrant)

The documentation requirements for getting a topical antimicrobial product approved to market in this country are fairly straight-forward. For example, the product must be manufactured under Good Manufacturing Practices (GMPs), and laboratory testing of the product must be performed under Good Laboratory Practices (GLPs). If the product is considered a drug, a New Drug Application (NDA) or an Abbreviated New Drug Application (ANDA) must be filed. Regulatory agencies such as the FDA also require that a battery of specific tests be conducted. These include such things as clinical trials, time-kill, and minimum inhibitory concentration (MIC) studies.

3. Cultural Requirements (Lower Left Quadrant)

The products must be perceived as high quality. And this perception is based squarely in the "world view" of the culture, including shared values, beliefs, and goals, as well as those of any subcultures, when applicable.[6]

Cultural and subcultural "world views" are not easily perceived by those belonging to that culture or subculture. For example, the traditional member of our western, science-based culture may chuckle at New Age advocates who believe in such things as healing crystals, and membership in a larger eco-spiritual continuum. But members of the scientific community in this country often extend the domain of science to their operative belief system and are unaware of it. For example, read a statement by Candice Pert, an eminent neuroscientist who discovered the "opiate receptors" of neuropeptides responsible for the brain's production of the morphine-like substances, "endorphins," which create the euphoria ("high") associated with physical exercise, etc.

> Scientists by nature are not creatures who commonly seek out or enjoy the public spotlight. Our training predisposes us to avoid any kind of overt behavior that might encourage two-way communication with the masses. Instead, we are content to pursue our truth in windowless laboratories, accountable only to members of our exclusive club.[17]

There is no persuasive evidence that the paragraph above is based on scientific reason. The position has nothing to do with "science," but everything to do with the "world view" of Pert's subculture, the scientific community.

An example more relevant to topical antimicrobial product development would be product ingredients that are viewed as "artificial," when "natural" ingredients are more valued. In this case, acceptance of the product containing "artificial" ingredients is likely to be impaired. Also, social institutions within a culture—the family, religious groups, social groups, and political groups—are important and manufacturers should take care not to alienate them with "thoughtless" advertising. Shared beliefs and values concerning the firm that makes the product are critical. If a manufacturer is perceived by consumers as a quality, caring firm, it will be easier for it to introduce and sell new products than if it is perceived otherwise.

Much of social reality is constructed—made up—by cultures.[10] However, the degree to which reality can be a cultural construct is bound by objective reality.[3,5] What we believe to be true or real, therefore, is often only partially true or real.

Sociologists tell us that there are three stages in socially constructed reality: externalization, objectivation, and internalization.[6] Externalization is the initial stage where a theory or opinion is accepted "tentatively" as being true. In the next stage, objectivation, the theory or opinion is accepted as "fact." And, finally, internalization is when the "accepted facts" are incorporated into a person's psyche as "absolute" truth.

It is often easier to use socially constructed reality as an ally to marketing programs than to educate people concerning truth. For example, with household antimicrobial products, when consumers read an advertisement stating the product "kills 99.9% of all germs," they literally believe only a very small number of disease-causing microorganisms survive. Yet, in truth, an error has been committed—that of applying a linear "proportion" measurement (% reduction in microorganisms) to an exponential distribution. The method of calculating the percent $\log_{10}$ reduction uses a percent (linear) on an exponential distribution (non-linear). If a linear percent measurement was calculated on a "linearized" exponential distribution, the common interpretation would be correct and the % reduction would be far less than 99.9%.[18] But, for purposes of advertising, the illusory presentation sounds better to consumers who will buy according to their constructed reality.

4. Personal Subjective (Upper Left Quadrant)

Individual members of a group (e.g., surgical staff) need to find value and construct positive beliefs about a specific product, if it is to be successful. Therefore, their perspective is important. Individuals must believe that the product was designed with them in mind and providing them specific examples, comparisons, and test conclusions is a very effective way of externalizing, objectifying and, finally, internalizing the advertising. But one must be certain that the claims can be supported and are grounded in objective reality.[2,16,19] Recall that socially constructed reality is bounded in objective reality.[5] Hence, when people are told that a product is easy to use, they must find this to be so when they use the product or their beliefs will shift to match the reality of their experience.

D. Surgical Scrub Formulations

1. Personal Objective (Upper Right Quadrant)

Surgical scrub formulations are designed to remove both transient and normal (resident) microorganisms from the hand surfaces.[19] Surgical scrubs, to be effective, must demonstrate immediate, persistent, and residual antimicrobial properties and must be low in skin irritation effects when used repeatedly over a prolonged period.

The product's immediate antimicrobial efficacy is a quantitative measurement of both the mechanical removal and immediate chemical inactivation of microorganisms residing on the skin surface.[20] The persistent antimicrobial effectiveness is a quantitative measurement of the product's ability to prevent microbial recolonization of the skin surfaces, either by microbial inhibition or lethality. The residual efficacy is a measurement of the product's cumulative antimicrobial properties after it has been used repeatedly. That is, as the antimicrobial product is used over time, it is absorbed into the stratum corneum of the skin, and as a result, prevents microbial recolonization of the skin surfaces.

2.   Social Requirements (Lower Right Quadrant)

The documentation and legal requirements for surgical scrub products are similar to those for the preoperative skin preparations. In general, products must meet the efficacy requirements of clinical trials utilizing human test subjects, as well as a series of in vitro tests, including time-kill and minimum inhibitory concentration studies. The actual requirements are presented in the FDA's Tentative Final Monograph for OTC products (21 CFR, Parts 333 and 339).

3.   Cultural Requirements (Lower Left Quadrant)

In general, these are the same as those presented in the discussion of cultural aspects relating to preoperative skin-prepping formulations.

4.   Personal Subjective (Upper Left Quadrant)

These issues are covered in the preoperative skin preparation portion of this work. Because surgical staff members actually use the product on themselves, however, its aesthetic attributes tend to be more important than for preoperative skin-prepping formulations.

It is important that relevant, but unknown product attributes be identified on the basis of evaluations by the potential product users. This will include subjective testing to evaluate the sensory attributes of the product—its container, its packaging, and other aesthetic concerns—in order to engineer a more desirable product.[7] Some of the areas of interest are presented in Table 1.

Developing a list of attributes from which panelists may select is difficult. One effective way of doing so is through the use of focus groups.[7] These groups consist of small numbers of surgical staff members (5–10) literally sitting down together and coming up with attributes of a surgical scrub product important to them. Once the subjective characteristics deemed important for success of a

Table 1 Areas of Evaluation of the Sensory Attributes of a Product

Sensory evaluation	Acceptance attributes
1. Clear	1. Like appearance
2. Opaque	2. Like fragrance
3. Strong smell	3. Like texture
4. Light smell	4. Like feel after wash
5. No smell	5. Like overall
6. Oily feel	6. Purchase intent
7. Dry feel	
8. Soft feel	
9. Lathers well	
Performance (nonantimicrobial)	Image
1. Lather well	1. Effective
2. Feels good on hands	2. Unique
3. Does not irritate hands	3. Good for hands
4. Conditions hands	4. Won't ruin gloves
5. Removes oil from hands	5. Won't irritate hands
6. Feels clean	6. High quality product

surgical scrub product have been determined, it is important to actually perform preference-testing by enrolling surgical staff personnel in a study and having them evaluate several configurations of a manufacturers surgical scrub product, as well as those of the competition.

E. Healthcare Personnel Handwash Formulations

1. Personal Objective (Upper Right Quadrant)

Healthcare personnel handwash formulations are intended to quickly remove any transient, pathogenic microorganisms picked up on the hands of a healthcare provider from patient A and prevent their passage to patient B. Hence, the product is intended to break the disease cycle at the level of the contaminated healthcare worker's hands by removing potentially infectious microorganisms.

The product must demonstrate low skin irritation upon repeated and prolonged use—20 to 30 washes per day for five consecutive days. Mildness to the hands, however, is usually attained at the price of reduced antimicrobial effectiveness. An optimal formulation which provides substantive reductions in contaminative microorganisms, yet is gentle to the hands, is the objective of product development.

Mildness can generally be built into the handwash in three ways:[20]

1. By proportionally reducing the irritating active ingredients such as chlorhexidine gluconate, iodophors or alcohol—for example, instead of the customary 4% level of chlorhexidine gluconate (CHG) found in surgical scrub formulations, a 1% or 2% chlorhexidine gluconate formulation may be developed.
2. By adding skin conditioners or emollients—these tend to counteract the irritating effect of antimicrobially active compounds, making the product more gentle and mild to skin.
3. By using a combination of these two methods (i.e., a reduction in the levels of active ingredient and the addition of emollients and skin conditioners—since so many healthcare personnel products were originally marketed as (much harsher) surgical scrubs, the market is vulnerable to a manufacturer or customer who will develop or select a product appropriately designed for its intended application.

2. Social Requirements (Lower Right Quadrant)

The documentation and regulations required to market a healthcare personnel handwash product are covered in the discussion of social requirements for pre-operative-prepping products.

3. Cultural Requirements (Lower Left Quadrant)

The healthcare personnel handwash product must be perceived as a highly effective antimicrobial compound, capable of removing and/or killing the microorganisms with which healthcare personnel may become contaminated.[20] Important shared beliefs/values relevant to healthcare personnel handwash formulations include its ability to inactivate certain pathogenic microorganism species perceived as being important indices of the product's effectiveness. The perception of importance for many of these is not grounded in objective fact, but, nevertheless, they remain "important" because they are "believed" to be important. For example, the anaerobic bacterial species, *Clostridium difficile*, will not grow in the presence of free atmospheric oxygen. However, the species is capable of producing a spore, and most healthcare personnel handwash formulations are not sporicidal. Many healthcare workers believe that formulations should be able to kill *C. difficile* spores, if the products are to be used by healthcare personnel. Although contaminative nosocomial disease potential for *C. difficile* is almost non-existent, but the perceived value of demonstrating product effectiveness against *C. difficile* is great.

Another aspect which is perceived as "valuable" in healthcare personnel handwash products is mildness to the hands. The product must not irritate the users' hands or create a general impression of harshness. Most healthcare personnel—particularly physicians—are very conscious of their hand-skin integrity.

They do not and will not use products that they perceive as being harsh. Some of the important attributes that must be determined relating to product quality include those presented in Table 1; hence, a focus group of 5 to 10 healthcare personnel should meet in focused forums to determine qualities of importance which must be built into the product.

4. Personal Subjective (Upper Left Quadrant)

Once general subjective characteristics of importance to the healthcare personnel have been determined, it is important that one actually perform preference testing.[7] The goal is to engineer a product which healthcare personnel prefer over those of the competition. This is done by enrolling a number of healthcare personnel as panelists to provide subjective evaluation of several configurations of test product and, preferably, even those of competing products. Other important considerations regarding personal subjective attributes have been discussed in the section relating to preoperative skin-prepping formulations.

F. Food Handler Handwash Formulations

1. Personal Objective (Upper Right Quadrant)

The potential for food handlers to be vectors in the transmission of food-borne disease is very significant.[21,22] Contaminating microorganisms are responsible for outbreaks of infectious diseases passed from food handlers to consumers via the food they eat. One of the most common sources of this is food handlers contaminated with enteric microorganisms from hand contact with their own feces. A significant problem in the food handling arena is that many who handle food do not wash their hands after defecation. To compensate for this, many food establishments have required that food handling personnel wear barrier gloves.[21] A vinyl or latex barrier glove which is intact (has no holes, rips or punctures) and uncontaminated will undisputedly provide protection from microbe transmission. But vinyl food-grade gloves, those most frequently used in the food service industry, commonly have pre-existing pinhole punctures which compromise the barrier protection. Additionally, these gloves are easily ripped, torn, or punctured as personnel perform their normal duties and, in many cases, such damage remains unknown to the wearer. Exposure to heat also has been reported to alter the integrity of barrier gloves significantly, making them brittle and thus, prone to breakage. Hence, the actual protection provided by barrier gloves is often much less than assumed. For example, in a study conducted at this facility, the hands of volunteer human subjects were inoculated with a strain of *Escherichia coli*.[21] The subjects then donned vinyl food handler gloves, each of which had four small needle punctures. Within five minutes, the outside of the gloves was sampled for microbial contamination. The results of this testing demonstrated that significant numbers of *E. coli* can be transferred from contaminated hands onto the outer surfaces of the gloves, if even small holes exist in the gloves.

On the other hand, wearing gloves actually may serve to increase the potential for disease transmission. As one wears vinyl or latex gloves for an hour or so, the microorganism populations on the hands within increase dramatically because the gloves prevent aeration of the hands, thereby increasing the levels of moisture, nutrients, and various other factors necessary for the growth of microorganisms. This phenomenon has long been known in the medical field, where mandatory handwashes using antimicrobial soap are required prior to gloving. Logically, as population numbers of both resident and contaminating microorganisms increase, so does the potential for disease transmission. Hence relying solely on barrier gloves to prevent disease is not prudent.

Handwashing has been used for years to prevent food-borne illness.[13,22] Handwashing effectiveness is dependant on two factors: (1) the physical removal of microorganism and (2) the immediate inactivation of microorganisms through contact with an antimicrobial ingredient in the soap. But that is not all; many antimicrobial ingredients have the ability to prevent transient microbial recolonization of the hand surfaces after handwashing by either microbial inhibition or lethality. If a non-antimicrobial soap is used, only the mechanical removal of microorganisms is significant.

In general, handwashing is very effective in removing contaminating microorganisms, given the handwash is performed "correctly." But assuring that food handlers perform effective handwashes, or even wash their hands at all, is difficult. First, the hands of food handlers can be exposed to different soil loads. The hands of those who work with pork products and high fat content hamburger may be very greasy. Medium grease loads are usually encountered among food handlers working with beef and chicken. Finally, for those working with salads and vegetable products, the hands tend to not be greasy at all, but dry and chafed.

When food handling personnel work exclusively with any one of the three food categories mentioned above, the problem is more easily addressed. The use of a heavy degreaser for those having high fat content exposure, a medium degreaser for those exposed to medium grease loads, and products with no degreaser for no-grease exposure, but an extra supply of skin emollients to help retain moisture and oils in the skin. If a food handler must work in all grease level conditions, a medium degreaser soap will probably suffice, along with the use of skin conditioning/moisturizing creams between washes, as needed.

The antimicrobial activity of food handler products should be very high. This is because food handlers tend to perform washes less thoroughly than do healthcare personnel, often leaving dirt and grime under the fingernail beds, which then may serve as a contamination source.

2. Social Requirements (Lower Right Quadrant)

Currently, regulation of soap products created for use in the food industry is somewhat vague. The USDA's "E" rating system is obsolete, as the regulatory function has passed from the USDA to the FDA. Currently, no study design has been accepted by the FDA for use of human subjects to demonstrate efficacy of food handler handwash products. Many manufacturers have adopted a modification of the healthcare personal handwash using *Escherichia coli* as the contaminating microorganism species instead of *Serratia marcescens*, the contaminating microorganism most commonly used in those evaluations. And the toxicological properties of handwash ingredients is very important because residues may be transferred from the hands of food handlers to the food they handle.

3. Cultural Requirements (Lower Left Quadrant)

In general, these are similar to those discussed relative to healthcare personnel handwashes; a major difference, however, is the high degree to which medical personnel value clean hands. Foodhandlers often have no real shared values relating to clean hands. Handwashing is just something they must do while on the job. This is probably because food handler positions generally are not filled by professionals, but by individuals from lower socioeconomic levels, young people, and, often, unmotivated individuals.[21]

To a large degree, culturally shared values of food handlers will have to be instilled by the food industry itself. This process will include training (upper right quadrant aspect), which will stimulate a value in "doing it right" (upper left quadrant aspect). In addition, in order to reduce the potential for disease transmission from fecal contamination, the following four steps will be useful in creating shared values/beliefs for these workers and the industry.[13,21]

1. Both gloving and handwashing with an effective antimicrobial product should be required for those performing high-risk tasks such as handling, cooking or wrapping food. While neither practice is failsafe, it is probable that the combination will provide more protection against disease transmission than either used alone.

    As mentioned earlier, observations in testing performed in this laboratory showed that population numbers of contaminative *E. coli* actually increased on the gloved hands when glove changes were performed at one or three hour intervals. However, concurrent testing showed that a thorough handwash using an effective antimicrobial product prior to gloving prevented significant growth of the contaminative microbes on the hand surfaces over the course of three consecutive hours of wear. The obvious conclusion is that, before gloves are put on, a thorough handwash should be performed using an effective antimicrobial soap. Even so, when feasible, no direct hand/glove contact with food should occur, and sanitized serving tongs or other utensils should be used for its manipulation.

2. Mandatory, ongoing sanitation training and education should be required of all employees. This is particularly necessary with inexperienced and/or unmotivated workers. Emphatically, without the active participation of employees, achieving adequate sanitation standards will be very difficult.

3. A high degree of personal hygiene should be required of food service personnel. Uniforms should be clean and changed often, employees should bathe or shower often, and they should not perform high-risk tasks when they are ill. High-risk tasks include hand/glove contact with food.

4. A quality control program supervised by qualified personnel should be initiated at each food service facility to monitor handwash/gloving practices. Written standard operating procedures (SOPs) should be drafted, and all employees formally trained in handwashing/gloving procedures. Training should be documented in employees' training records, and handwashing procedures posted in a conspicuous place near sinks used for handwashing.

4. Personal Subjective (Upper Left Quadrant)

Preference testing will be extremely valuable to determine just what types of product attributes food handlers prefer.[14,20] The goal is two-fold: (1) to build a product that food handlers will want to use and (2) to build a product that is preferred over those of the competitors. Since relatively little is known concerning the subjective preferences of food handlers, it is important that such testing include enrolling a reasonably large number of food handlers in a study to evaluate various products sequentially. For example, begin with fragrance preferences. When one or two preferred fragrances are identified, add lathering characteristics to the evaluation. Then, with these attributes identified, evaluate the feel of the hands after washing. And this type of interactive process can be conducted for all other attributes deemed important.

G. Antimicrobial Handwash and Bodywash Formulations

1. Personal Objective (Upper Right Quadrant)

Whether these categories become officially acknowledged by the FDA probably will not matter, because consumers want antimicrobial hand and body soaps. The antimicrobial compounds generally used in these types of products are Parachlorometaxylenol (PCMX), triclosan, and isopropyl alcohol, all of which have been used in handwash products for many years. The antimicrobial handsoaps/ bodysoaps must also be of low irritation potential.

Perhaps the biggest problem with antimicrobial handsoaps is that the various claims some manufacturers present are misleading and even false. It is critical that manufacturers take the responsibility for manufacturing products which perform well both in clinical trials and in vitro sensitivity testing.

2. Social Requirements (Lower Right Quadrant)

Since FDA requirements do not exist for antimicrobial handsoaps, regulation of quality is left to the industry, particularly the CTFA and SDA. These organizations have truly taken responsibility for setting realistic standards to which members voluntarily conform. However, there continues to be concern with various manufacturers being unable to assure and verify their label and advertising claims to the satisfaction of the Federal Trade Commission.

3. Cultural Requirements (Lower Left Quadrant)

There is a tremendous need for consumers to feel clean. The shared values, due mainly to effective advertising campaigns, have convinced people that, if they use antimicrobial hand/body soaps, they will be cleaner than if they do not. And, as a bonus, if they use antimicrobial bodywashes, they will not offend others with body odor.

There is nothing wrong with cultural beliefs that antimicrobial soaps are better so long as manufacturers truly strive to meet their stated labeling/advertising claims. That is, they must support their claims with data collected from valid, statistically complete studies using human subjects, as well as from in vitro testing to assess microorganism sensitivity to the product.

4. Personal Subjective (Upper Left Quadrant)

Soap manufacturers have mounted a very strong campaign for antimicrobial soaps. And, frequently, manufacturers of these consumer products have really taken the time to understand the personal subjective and cultural value, and have used that knowledge to launch highly successful marketing campaigns.

III. CONCLUSION

It is important that, in developing topical antimicrobial wash products, a complete, multidimensional approach be taken. This will help ensure that the product which results is designed for the specific needs of the market and that those needs are met. In this way, the product is more likely to have a long, useful, and profitable life.

2
Skin Properties

To better understand topical antimicrobial product evaluations, it is necessary to discuss the anatomical structure of the skin, or integument. Human skin, the largest single organ, provides a continuous covering over the body. The skin constitutes about 16 percent of the total body weight, having an area of approximately 1.8 m^2 for average-sized adults.[23]

Because skin is under slight tension, it is observed to shrink when a section is removed from the body. The surface structure of skin appears to contain innumerable etch lines and wrinkles that are derived from underlying fibers in the tissues. In most skin areas, the fibers are irregular so that the skin etch lines form a distinct "mosaic" pattern. However, on surfaces such as the finger and toe pads, and to a lesser extent the palms and soles, is a unique regular and consistent arrangement of skin ridges and furrows that are acquired before birth and retained throughout life.

More obvious skin etch lines and folds are seen at what are termed "skin joints." These are areas where the skin is bound firmly onto the deeper tissue layers. The skin at other parts of the body is loose, separated from muscles and organs by a layer of subcutaneous connective tissue which is interspersed with varying amounts of fatty tissue. In anatomical locations where the skin is relatively loose, it also tends to be elastic. However, this elasticity is gradually lost with age so that the skin becomes flabby and folded.

I. STRUCTURE

The skin consists of two general layers, the epidermis and the dermis (Figure 2), which are composed of cells in varying stages of keratinization.[24] Both layers contain elastic fibers that largely determine the elasticity and firmness of the skin. The epidermis contains no blood or lymph vessels. However, the inner layers consist of metabolically active cells, strongly bound together by spot desmosome junctions. Underlying the epidermis is the dermis, which is made

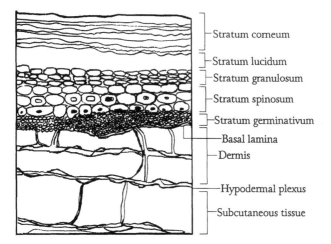

Stratum corneum

Stratum lucidum

Stratum granulosum

Stratum spinosum

Stratum germinativum

Basal lamina

Dermis

Hypodermal plexus

Subcutaneous tissue

Figure 2 Cross-section of the composition of human skin.

up of connective tissue containing a rich supply of both blood and lymphatic vessels, and varies in thickness in different regions of the body. Human dermal skin comprises two layers with rather indistinct boundaries: the outer papillary layer and the deeper reticular layer. The papillary layer is thin and composed mainly of loose, connective tissue. This layer contains fibroblasts, mast cells, and macrophages, as well as other migrant leukocytes. The reticular layer is thicker than the papillary layer and is composed of dense, irregularly arranged, connective tissue. It contains more collagen fibers and fewer cells than the papillary layer.

The subcutaneous tissues lying beneath the dermis vary in thickness, as well as histological makeup in different anatomical regions and contain a vascular network which supplies nutritive, immunological, and coagulative properties. Additionally, a nerve network connected with various subcutaneous fibers brings sensory stimuli from the skin surface to the brain and conveys certain autonomic motor stimuli from the brain to various specialized involuntary muscle structures.

Let us now look at the epidermis and dermis in greater detail.

II. THE EPIDERMIS

The epidermis, the outer layer of the skin, has a thickness varying from 75 to 150 μm. It normally presents an impenetrable barrier to microorganisms, while also providing a suitable environment for a variety of "resident" microorganisms. These residents serve to impede epidermal colonization by other poten-tially pathogenic microorganisms.[23,24]

The epidermis comprises several layers than can be visually distinguished from each other. The inner-most layers consist of metabolically active cells strongly bonded by spot desmosome junctions. As can be seen in Figure 2, the epidermis is stratified and lies on top of the dermis region. It is separated from the dermis by the basement membrane, or basal lamina, a continuous and struc-tureless sheet distinct from the cytoplasmic membranes of the adjacent cells. It is supported by numerous extracellular fibrils, which can be regarded as part of the basal lamina, as well as by some of the uppermost dermal cells.[25,26]

The basal lamina supports a layer of basal cells which forms the stratum germinativum (or stratum basale) of the epidermis. These cells appear to lack continuous attachment to one another, but are connected to each other by many "finger-like" projections from the upper epidermal layers and contain many dense particles consisting of ribonucleoprotein. The cells are metabolically ac-tive, but histologically undifferentiated at the deepest level. The stratum germi-nativum is seen to consist of several cell layers in advancing stages of cellular differentiation until they have moved toward the surface to compose the *stratum spinosum* cell layer. The stratum spinosum consists of irregular-shaped "prickle" cells (relatively large cells containing large numbers of keratin filaments) which become increasingly elongated and flattened as they move outward toward the skin surface. These cells are firmly bound together by cytoplasmic expansions filled with bundles of filaments called tonofibrils. Desmosomes cover the cell surfaces, giving the cells a prickly, studded appearance in the light microscope. The progressive cellular differentiation is accompanied by the production of a low density intercellular "cement" that augments the cellular adhesive proper-ties, assuring that a firm, continuous, tough layer is formed. Intercellular bridg-ing structures are also present and assist in the formation of a compact tissue. The term, *Malpighian layer*, denotes the *strata germinativum* and *spinosum* con-sidered together.[25]

Superior to the stratum spinosum lies the stratum granulosum. This layer consists of one to several layers of granular cells which are usually morphomet-rically larger than cells found in the stratum spinosum and are nucleated, having cytoplasms containing numerous aggregates of poorly characterized material called keratohyalin, as well as tonofibrils which appear fragmented. Both kerato-hyalin and tonofibrils play a role in the process of skin keratinization. The gran-ular cells are in the penultimate stages of keratinization.

The layers of the epidermis residing above the stratum granulosum consist of dying and dead cells in various stages of keratinization. The deepest of these keratinized layers, the stratum lucidum, is present only in areas of thickened skin such as the palms and soles of the feet. The cells of the stratum lucidum appear translucent and extremely flattened, containing only disintegrating cell nuclei, and keratohyalin material not visually distinguishable with a light micro-scope.[24]

The stratum corneum, the most external component of the epidermis, con-sists of several layers of flattened, dead squamous cells containing large amounts of keratin and firmly attached to one another. The intercellular spaces are filled with multiple lipid bilayers, packed so that there is an alternating pattern of nonpolar hydrocarbon regions and polar head group regions. The lipids involved are cermides, cholesterol, cholesterol sulphates, and free fatty acids organized together in multilaminated shells. The bilayers consist of straight, loosely-packed, saturated hydrocarbon chains, without structural peru-bation among the hydrophobic chains. This highly ordered, rigid structure re-sults in a unique impermeability for many compounds, including water. In some cells, remnants of nuclear membranes can be observed with a light microscope. The most external cells of the stratum corneum are constantly shed as minute skin particles (squames) and, in certain conditions, as visible flakes and sheets (e.g., dandruff). This cell loss is compensated by constant replacement of cells rising from the lower epidermal layers. That is, epidermal cells continually move toward the skin surface in successive stages of cellular differentiation and finally death, until they are lost to the environment via exfoliation. This exfoliative process provides a constant replacement of nutrients for resident microorgan-isms colonizing the skin surface. For the most part, the nutrients present on the skin surfaces are insufficient to support growth of fastidious, nonresident microorganisms.[27]

The epidermis varies in thickness according to anatomical region and gen-der. It is slightly thicker on the dorsal and extensor aspects than on the ventral and flexor aspects of the body and, as a rule, is thicker in males than in females. It also changes in thickness with age; an infant's skin is thinner than an adult's, but as the individual continues to age, the skin thickness declines, and it be-comes increasingly fragile.[24]

The regeneration of the epidermis begins when a basal cell divides, result-ing in occupation of two basal positions by the daughter cells. The next step is a migration into the prickle cell layer. The cells flatten, become granulated, and loose their organelles and nuclei as they move outward through successive strata to become keratinized squamous cells (corneocytes), finally flaking off from the skin surface.

The different layers of the epidermis all contain keratin, but during the process of keratinization, different types of keratin are produced. These are com-posed of 19 different α-helix proteins with molecular weights ranging from 60,000 to 68,000.[23]

In human skin, the normal epidermal cell cycle is 20–30 days in duration, depending on the region of the body. The regeneration of the epidermis is regu-lated according to the thickness of the epidermis. Faulty control of the rate of proliferation causes a skin disorder called psoriasis in which the rate of basal cell proliferation is greatly increased, and epidermal cell cycles are completed within a week, without complete keratinization.

Interspersed with the cells destined for keratinization, the so-called keratinocytes, the epidermis contains small numbers of macrophage-like Langerhans cells, melanin-producing melanocytes, and neural Mercell cells.[28]

III. THE DERMIS

The dermis, unlike the epidermis which overlays it, contains blood vessels, an intercellular matrix, and fluid and lymphatic vessels that provide the dermal cells nutrients and immunological protection.[25] The human dermis consists of two layers, the boundaries of which are rather indistinct—the outermost papillary layer and the deeper reticular layer. The papillary layer is thin and composed of loose connective tissue, fibroblasts, mast cells, macrophages, and extravasated leukocytes. The papillary layer penetrates into the papillae. Collagen fibers from the papillary region extend through the basal lamina and into the epidermis. They are thought to have the special function of binding the dermis to the epidermis and are often kernatinal anchoring fibers. Intercellular substances provide strength and support of tissue and act as a medium for the diffusion of nutrients and metabolites between blood capillaries and the dermal cells in support of cellular metabolism. The papillary layer adheres tightly to the basement membrane of the epidermis, while the lower surface merges gradually with the reticular layer.

The reticular layer is thicker than the papillary layer and contains varying amounts of fat, often as a function of an individual's sex, age, and anatomical region of the body. Females generally show a greater amount of fat within the subcutaneous layers than do males. The reticular layer also contains the larger blood vessels and nerves, from which arise various superficial cutaneous branches to supply the skin.[28,29]

Both layers of the dermis are interspersed with connective tissue consisting of bundles of collagenous, elastic, and reticular fibers grounded in an amorphous, intercellular substance containing various types of histologically discrete cells. The dermal fibers add strength and elasticity to the skin, as well as account for its patterning and cleavage lines. In the papillary region, the bundles of collagenous and elastic fibers tend to be widely separated. In the reticular layer beneath, however, lies a dense network of coarse collagenous fiber bundles situated more or less parallel to the skin surface.

At least three important, histologically distinct cell types are found in the dermis.[24,25,28] The first, the fibroblast, is an undifferentiated cell capable of transforming into other connective tissue cell types. Another is the histiocyte, a cell capable of phagocytosing inorganic or organic particles, including bacteria and fungi. Histiocytes have been observed to unite in certain situations to form macrophagous "giant" cells, which are able to phagocytose relatively large particles. The third important cell commonly found in the dermis is the Langerhans cell. Langerhans cells, morphologically, resemble the fibroblast and are observed in the upper papillary layer, especially around capillaries, where they are often observed to be arranged in concentric rings.

The dermis is normally protected by the overlying, tough, keratinized epidermal covering which prevents the entry of pathogenic microorganisms. However, when the epidermis is compromised, and microorganisms penetrate to the dermis or beyond, they may elicit both inflammatory and immune responses in normal, healthy individuals.

IV. DERMAL VASCULARIZATION

An important component of the dermis is its vascular system, which is composed of a rich network of blood and lymph vessels. Arterial blood brings oxygen and nutrients to all dermal cells that are utilized at capillary junctions. The veinous blood carries away products of metabolism collected at the capillary junctions. Additionally, the vascular system carries cells of the immune system to all parts of the dermis.[23,24,25]

Figure 3 presents a diagrammatic representation of the various cell structures found in normal skin tissue. Notice that the skin is supplied by small arteries (arterioles) embedded in the subcutaneous layers. The arterioles form branches that pass outward and form the hypodermal arterial plexus. Branches from this plexus provide the vascular supply of hair follicles and sweat glands. Others pass through the dermis and branch to form dermal and subepidermal plexuses which give rise to many short vessels. The subepidermal plexuses, that lie immediately below the basement membrane of the epidermis, form the arterial arms of the capillary loops which project into the dermal regions. The blood passes through the arterial capillary loops into the wider venous arms and collects in a series of venous plexuses embedded at various levels in the dermis. Finally, blood reaches the hypodermal venules and passes thence to the subcutaneous veins. The arterial network is especially rich in the palms, soles, and the face, the vessels being both larger and more numerous than elsewhere. The cutaneous/cartilaginous appendages (e.g., ears, nose) also have rich vascular supplies.

Because the capillary loops have very small diameters, approximately that of a red blood cell, temporary blockage of the capillary may occur. Further, the flow of blood through the capillary field is intermittent and may stop and start. There is also a constant variation in the number of capillaries open at any one

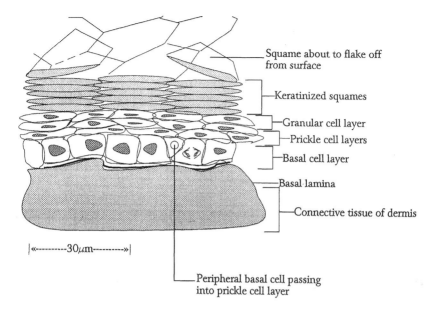

Figure 3 Various cell structures found in normal skin tissue.

time, since the opening and closing of capillaries is one of the mechanisms by which the body controls heat loss. Despite the numerous plexuses supplying the skin, the vascular surface available for the exchange of oxygen and nutrients between the blood and the tissues is smaller than that of organs such as muscles. The dermis contains various structures of epidermal origin, namely hair follicles and sweat and sebaceous glands, which are more or less numerous, depending on the region of the body. A rich supply of nerves is also found throughout the dermis.[23,24,25]

A. Subcutaneous Tissue

This layer consists of diffuse connective tissue that binds the skin loosely to the subjacent organs. The overlying hypodermis often contains fat cells, varying in number and in size as a function of the area of the body and nutritional state of the individual.

B. Hairs

Hair shafts are thin, keratinized structures derived from invagination of the epidermal epithelium. Their color, size and disposition vary according to race, age, sex, and region of the body. Hairs grow discontinuously and exhibit periods of growth followed by periods of rest.

C. Sebaceous Glands

Sebaceous glands are found embedded in the dermis over all areas of the body, except in those areas lacking hairs. These glands open into short ducts which end in the upper portion of a hair follicle. Sebaceous glands arise structurally from undifferentiated, flattened epithelial cells. The cell nuclei gradually shrink, and the cells become filled with fat droplets and burst. As these cells differentiate, sebaceous glands move toward the surface of the skin. The sebaceous gland secretes sebum, together with remnants of dead cells. Sebum is composed of a complex mixture of lipids containing triglycerides, free fatty acids, cholesterol and cholesteryl esters and is secreted continuously. A disturbance in the normal secretion and flow of sebum is one of the reasons for the development of acne.[27]

D. Sweat Glands

The sweat glands are widely distributed in the skin in the form of simple, coiled, tubular structures. The fluid secreted by these glands contains mainly water, sodium chloride, urea, ammonia, uric acid, and proteins. The constant evaporation of sweat requires heat, which is withdrawn from the capillaries surrounding the gland; this loss of heat contributes to the thermoregulation of the body.[28,29]

V. THE TISSUE FLUIDS

Because there is no vascular network residing in the epidermis, nutrients and oxygen for epidermal cells must diffuse varying distances from the blood capillaries. Although the capillaries do not appear to be actively contractile, the endothelial cells of which they are composed have an elastic property that makes them resistant to distortion, so that changes in tone result in changes in capillary permeability. Under normal conditions of intracapillary pressure, there is a tendency for fluid to flow from the capillaries into the tissue spaces on the arterial side of the capillary loop and to be reabsorbed from the tissues on the venous side, thereby providing a constant circulation. The action of the muscular walls of the feeder arterioles and of the precapillary sphincters can produce great variations in capillary pressures.[24]

3
Skin Microbiology

Bacteria, fungi, viruses, and other parasites are the causative agents of microbial disease. However, in evaluating topical antimicrobial products such as surgical scrubs, pre-operative prepping solutions, and health care personnel handwashes, bacteria are the indices used in estimating antimicrobial effectiveness.

I. ETIOLOGY OF INFECTIOUS DISEASES

For an infectious disease to spread, the following events need to occur:[30]

1. Encounter: The host must be exposed to the microorganism.
2. Entry: The microorganism must enter the host.
3. Spread: The microorganism must spread from the entry site.
4. Multiplication: The microorganism must multiply in the host.
5. Damage: The microorganism and its metabolytes and/or the host's immunological response causes tissue damage.

All of these steps are required in breaching the host's defenses. The actual disease presentation produced by the microorganism depends upon the entry conditions, the host's defenses, and the species and strain of microorganism.[31]

A. Encounter and Entry

The host is exposed to the microorganism in a variety of situations relevant to topical antimicrobial compounds.[32] Generally, there is (1) the exposure of patients to pathogenic microorganisms via a healthcare/foodhandling worker's contamination by a different patient, worker, or environmental source; (2) the exposure of a patient's surgical site or a worker's wound site to contaminated hands; and (3) the contamination of a surgical, catheter or injection site with the patient's own normal microbial flora.

In Situation #1, the patient is exposed to pathogenic microorganisms from a different patient via the hands of contaminated healthcare personnel.[33] (This is also the situation of a food-borne disease but it will be discussed in a separate section of this book.) That is, the healthcare personnel serve as a disease vector, transferring pathogenic microorganisms from patient A to patient B. It should be noted that the microorganisms transferred are transient pathogens, not normal microorganisms of the skin.

Situation #2 represents the exposure of a patient's surgical site by the contaminated hands of surgical personnel. This generally occurs when a surgical glove is torn or nicked, and the normal resident microorganisms of the caretaker's skin enter the patient at the surgical site.[33] The degree of infectious disease is dependent upon the surgical site, the number of opportunistic microorganisms transferred to the patient, and the immunological competence of the patient.

Situation #3 occurs as a result of the contamination of a surgical site, venous/arterial catheter site, or an injection site with the patient's normal skin microorganisms.[33,34] Contaminative (nosocomial) infections of surgical and catheter sites are not uncommon and are dangerous because of entry of microorganisms into the blood stream, resulting in a bacteremia and septicemia. Patients who are immunocompromised are at a greater risk of septicemia. Injection site infections are generally localized, but can result in a septic condition.

B. Spread

Spread of pathogenic microorganisms from patient to patient via healthcare personnel hand contact can take a variety of forms. It can be hand-to-hand contact, such as shaking hands. Often, the contaminating bacteria are respiratory or gastrointestinal in origin, but not exclusively. Most anatomical sites subsequently examined, particularly if they are anatomically or physiologically compromised, can become a spread site.[30,35]

General infection resulting from direct contact with the hands of surgical personnel or introduction of normal skin microorganisms through catheters is facilitated by the blood and the lymphatic systems. Spreading can be passive or active, depending upon a microorganisms' motility and/or manufacture of extracellular hydrolytic enzymes that allow them to break through "walling off" mechanisms of the inflammatory response.[30] For example, *streptococci* bacteria produce a protease that breaks up fibrin, a hyaluronidase that hydrolyzes hyaluronic acid (an important component of connective tissue), and a deoxyribonuclease that causes the release of DNA from lysed white cells and reduces the viscosity of pus.[36]

C. Multiplication

Once the microorganisms have spread from the entry site, they frequently multiply to cause a systemic infection.[32,37] Generally, there is a time lapse between the exposure and the manifestation of symptoms, referred to as the incubation time (Figure 4).

The ability of pathogenic microorganisms to multiply in the body is influenced by the body's immunological system.

D. Immune System

In normally functioning immune systems, the body is able to generate a variety of "immune system" cells and antibody molecules which are capable of detecting ("recognizing") and eliminating an apparently infinite variety of foreign substrates and microbial forms, that include viruses, bacteria, and fungi.[38,39]

Functionally, the immune system can be divided into two interrelated activities: recognition and response. Immunological recognition is so highly specific that it is able to distinguish one foreign pathogenic microorganism from another. The immune system is also able to discriminate between "self" and "not self" at the molecular level of cell structure. Once a foreign microorganism or molecular form is recognized, the immune system mounts an immune ("effector") response through the participation of various immunological cells, phagocytic cells, and certain molecular substrates.[40] The effector response may then eliminate or neutralize the infectious microorganism species or foreign molecular form. The immune system is able to translate the recognition established at initial exposure to a foreign microorganism or molecule to subsequent exposures because certain immunoactive cells retain "memory" of them. Upon a subsequent exposure, a heightened immune response is generated which serves to eliminate the invader rapidly and even prevent a disease.

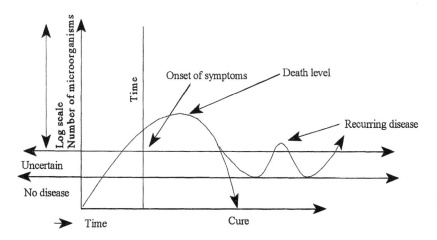

Figure 4 Incubation time.

An individual who is sick or a patient who has undergone a surgical procedure is often immunologically compromised.[41,42] Hence, the probability of morbidity and mortality through infectious disease is increased.

E. Microbial Nutrition

At first glance, it would appear that the body offers a variety of rich mediums for the support of microorganisms. Body fluids such as plasma contain sugars, vitamins, minerals, and other substances on which bacteria and fungi can subsist. For microorganisms other than normal flora, life in or on the body is not so simple. For example, if fresh blood plasma is incubated with challenge microorganisms, microbial growth is generally nonexistent or sparse. This is because antimicrobial substances such as lysozyme and molecular constituents of the complement system are inhibitory.[43,44]

Additionally, bacteria require free iron for the synthesis of their cytochromes and other enzymes. This appears to be a limiting factor for growth of bacteria in the body. Plasma, as well as a variety of other body fluids, contain very little free iron, probably due to its avid binding to a wide range of proteins.[45] In fact, the body actually sequesters iron to defend itself against bacterial multiplication.[30] When a sufficient number of microorganisms has been detected within the body, iron-binding proteins literally pour into plasma and other tissue fluids, as the body strives to limit the availability of free iron to bacteria.

The range of nutritional requirements of microorganisms that are found as normal microbial flora is a reflection of their ecological niches. For example, *Staphylococcus epidermidis*, the predominant skin surface-colonizing bacterium, requires several amino acids and vitamins that are commonly found on the skin surfaces.[30] However, microorganisms common in both soil and water are much less fastidious. They can achieve their organic requirements from simple carbon compound sources which are widely available in the body as well as the natural environment. Both *Escherichia coli* and *Pseudomonas* spp. are examples of bacteria that can thrive on very nutrient-minimal media.[46]

F. Physical Factors

Physical factors also affect microbial multiplication within the body. Important physical factors include temperature range of an anatomical region of the body, osmotic pressure of fluids and humidity.[47] Microorganisms that are normal inhabitants of the body, or those which can only live on the body, tend to have much more limited tolerance to physical changes than ones also found commonly in the environment, like *Pseudomonas sp.*[35]

G. Endogenous Microorganisms of the Body

It is important, in studying the role of topical antimicrobial products, to be familiar with the microorganisms that compose the normal flora. These are important, particularly in the surgical environment where normal flora species may be introduced into a surgical wound by surgical staff or contaminatively from the patient's own skin and opportunistically to produce an infection.

Normally, the body contains dozens of different bacterial species as well as numerous different viruses and fungi.[31] The most common sites of normal microbial colonization include the digestive tract (mouth and large intestine), the respiratory tract (nose and oropharynx), the female genitalia, and the skin surfaces.[44]

II. THE BACTERIAL CELL

In life forms, two basic cell types are found: eucaryotes and procaryotes.[35,47,48,49] Eukaryotic cells are unit structures of higher plants, animals, fungi, most kinds of algae, and certain single-celled organisms such as protozoa. Eukaryote cells are characterized by true nuclear bodies (nucleus) bounded by a nuclear membrane and containing chromosomes which undergo miotic division (and meiotic division in germinal cells).[35] They also possess various cellular organelles, such as Golgi bodies and the endoplasmic reticulum that function in sorting and transporting proteins destined for secretion or incorporation into the plasma membrane or lysosomes; mitochondria and chloroplasts (plants), both of which play critical roles in energy metabolism; and ribosomes, structures responsible for generating peptides and proteins.[50]

Eucaryotic cells have another level of internal organization: the cytoskeleton.[50] The cytoskeleton is a network of protein filaments extending throughout the cytoplasm to provide the structural framework of the cell, determine its shape, and generally organize the cytoplasm. The cytoskeleton is responsible for movements of the entire cell, such as in muscle contraction, as well as intracellular transport, and positioning of the organelles and other structures, including the movements of the chromosomes during cell division.[35,47]

Procaryote cells, which are simpler structurally than the eucaryotes, are the form of all types of bacteria and blue-green algae. They do not contain a nucleus bounded within a membrane but, instead have a simple nuclear filament, replicated nonmitotically.[50,51]

Procaryotic cells are typically enclosed in a rigid cell wall containing a unique constituent, muramic acid. Bacteria are classified morphologically into one of three categories: cocci (spheres), bacilli (rods), or spirals.[52] The cocci and bacilli are the most common.[51] Bacteria are distinguished further as being Gram-positive or Gram-negative, using the Gram staining method.[51] Devised by Hans Gram, the Gram stain ranks among the most useful and important stains for identifying bacteria, allowing one to distinguish between bacteria which exhibit similar morphology.[44] Gram-positive bacteria, following a destaining step in the procedure, retain the creystal violet-iodine complex and appear purple.[51] Other bacteria do not retain the complex when destained with acetone-alcohol and are subsequently stained red by the safrinen dye counterstain which follows. Using this procedure, one can determine in the microscope the bacterial size, form, and Gram-positive or Gram-negative property.* There are a number of other staining techniques useful in the study of bacteria, including acid-fast staining, capsule-staining, flagella-staining, metachromatic granule-staining, spore-staining, and relief-staining.[51]

While the great majority of bacteria fall into one of three basic morphologies (coccus, bacillus, spirals), bacteria may exhibit variations on or blending of shapes in older cultures, where bacterial cell structures become weakened and break down.[53] For example, cells may appear balloon-like, Y-shaped, or granular. Such degenerative forms result from the breakdown of mechanisms that control selective permeability and from enzymatic autolysis. Aberrant bacterial forms may also be produced by the culture of bacterial colonies under adverse environmental conditions, including higher than optimal temperature ranges, the presence of high concentrations of inorganic salts or exposure to sublethal levels of antimicrobial products.

A. Bacterial Cell Structures

Most bacteria are enclosed in multilayered cell structures, that include a cytoplasmic (plasma) membrane, a cell wall with associated proteins and polysaccharides, and in some, protective capsules and slime layers that help protect against the host's immune system, particularly phagocytosis.[30,47,50] Many also bear external filamentous appendages (flagella and pili) that function in locomotion, tissue attachment, and/or specialized reproductive processes.

Generally speaking, the bacterial cell wall is a rigid structure that encloses and protects the inner "protoplast" from physical damage (e.g., conditions of low external osmotic pressure). The rigid cell wall is thought to be an evolved structure that allows bacteria to tolerate a wide range of environmental condi-

---

*The difference between Gram-negative and Gram-positive can be attributed to differing chemical constituents in the cell wall. Gram-negative bacterial cell walls have a higher lipid content than do Gram-positive bacteria. Although a crystal violent-iodine complex is formed in both kinds of cell, the alcohol removes the lipid from the Gram-negative bacterial cells, which increases the permeability, resulting in loss of the dye complex. The complex is retained, however, in the Gram-positive cell walls, in which dehydration by the alcohol causes a decrease in permeability.

tions. The protoplast, a bacterium less the cell wall, is composed of a cytoplasmic membrane and its internal contents.

Internally, bacteria are relatively simple procaryotic cells.[51,52] Major cytoplasmic structures include a central fibrillar chromatin network surrounded by an amorphous cytoplasm containing ribosomes. Cytoplasmic inclusion bodies— energy storage granules—vary in chemical composition, depending upon the bacterial species, and in number, dependent upon the bacterial growth phase and environmental conditions (e.g., temperature). Some cytoplasmic structures (e.g., endospores) are limited to only a few bacterial species. Typical Gram-positive and Gram-negative bacterial cells, which differ primarily in cell wall organization, are shown in Figure 5.

B. Bacterial Size and Form

Medically important bacterial species vary from approximately 0.4 to 2 μm in size and appear under the light microscope as spheres (cocci), rods (bacilli), or spirals (vibros or spirochetes).[35,47] Cocci are found singly, in pairs as diplococci, in chains (e.g., *Streptococcus* spp.) or, depending upon division planes, in tetrads or in grape-like clusters (e.g., *Staphylococcus* spp.). Bacilli vary considerably in length and width, from very short rods (coccobacilli) to long rods that can measure several times their diameter. The terminal ends of bacilli may be gently rounded, as in enteric organisms such as *Salmonella typhi*, or squared, as in *Bacillus anthracis*. Long sequences of bacilli that have not separated into single cells are called chains, while long, thread-like bacilli are generally referred to as filaments. Fusiform bacilli, found in the oral and gut cavities, are tapered at both ends.[50] Curved bacterial rods vary from small, comma-shaped or mildly helical forms with only a single curve, such as *Vibrio cholerae*, to longer spiral (spirochete), multiply-coiled forms, such as species of *Borrelia*, *Treponema*, and *Leptospira*.

C. Cell Envelope

Bacterial cells are generally bounded by the cell envelope, an integrated structure of varying complexity and generally of several layers.[35,47,53] The bacterial cell envelope commonly includes a cytoplasmic membrane, an overlying cell wall, specialized proteins or polysaccharides, and various outer adherent materials. This multilayered structure constitutes 20% or more of the procaryotic cell's dry weight.[48] The bacterial cell envelope contains transport sites for nutrients, receptor sites for specific bacterial viruses, sites of host antibody and complement reactions and, components often toxic to the host.

The occurrence of bacteria in shapes other than spheres (cocci) is evidence that envelope rigidity is sufficient to withstand environmental surface tension

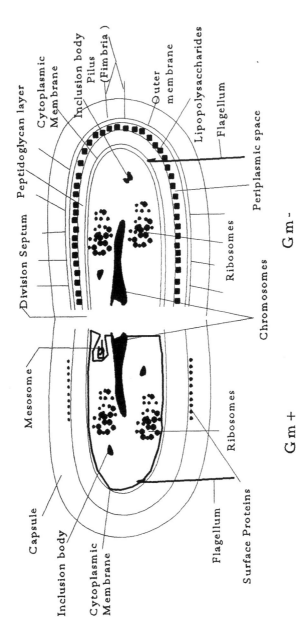

Figure 5 Cross-sections of Gram-positive and Gram-negative cells.

and internal turgor of the cell. This rigidity and the resulting cell shape are largely attributable to the cell wall component, which also provides resistance to mechanical disruption. Both the cell wall and the underlying cytoplasmic membrane are demonstrable in electron micrographs of thin sections or freeze-fracture preparations.

Among bacteria, the nature of the cell envelope varies considerably in structural complexity and differentiation. The simplest envelopes are found among the members of the genus *Mycoplasm* which possess only a cell membrane, which averages about 7.5 nm thick.[50] Because the cell membrane structure has the dual role of "cell wall" and "cytoplasmic membrane," and does not contain the peptidoglycans found in true cell walls, mycoplasmas possess limited cellular strength.

The envelope structure of Gram-positive bacteria is more complex (Figure 5). In addition to an inner cytoplasmic membrane, having the appearance in cross-section of a "double-track," or two-rail fence membrane, when viewed under the electron microscope, the cell is surrounded by an amphorous cell wall. The cell wall, viewed sectionally, does not exhibit the layered appearance of the cytoplasmic membrane and generally is from 15–50 nm in thickness.

The cell envelope of Gram-negative bacteria is the most complex (Figure 5). The inner layer of the envelope is a cytoplasmic membrane, structurally similar to that of Gram-positive cells. The overlying cell wall is composed of a second (outer) membrane often appearing wrinkled or with a "raisin-like" appearance.[52] Closely opposed to the inner surface of this membrane, not often visible as a separate layer, is a thick peptidoglycan layer. The Gram-negative wall is thinner than that of Gram-positive species, usually averaging about 10–15 nm across.

D. Capsules and Slime Layers

The virulence—the ability to cause disease—of certain pathogenic bacteria is often related directly to their production of a capsule.[30,31,35] For example, virulent strains of *Streptococcus pneumonae* produce capsules that protect them from phagocytosis by neutrophils and macrophages. It has been demonstrated that loss of the capsule-forming ability produces a loss of virulence and increased ease of destruction by host phagocytes. While capsules and slime layers are both composed of similar "gels" that adhere to the outer cell wall, slime layers consisting of extracellular polymers are more easily "washed off" than capsules.[50]

Most bacteria, possibly all species found in the environment (wild types), are surrounded by a layer of the gelatinous, poorly-defined and poorly stainable material that has been termed the capsule, slime slayer, or glycocalx. The term, capsule, is generally applied to the material surrounding a single cell, while the slime layer is more commonly used to describe the matrix that envelopes a microcolony or group of bacterial cells. These distinctions are arbitrary.[50]

The capsule enclosing the cell walls of some bacteria has been thought by some to be simply the waste products of cellular metabolism and not of great importance to bacteria.[51] This view, for the most part, has been seriously challenged.[50] It is clear that capsules play a key role in microbial ecology, not solely in the host/parasite relationship, but also in a species' growth and survival in the natural environment. Bacteria found, as in nature, whether derived from water or soil or from the mucus, blood, or tissue of a host, almost always possess a capsule. It appears that only after an in-vitro cultivation in the laboratory setting are the capsules of many species lost or diminished.

The capsules surrounding individual bacterial cells or the slime layers surrounding microcolonies of bacteria serve several known purposes.[48,49,50] They protect bacteria from desiccation, bacteriophage penetration, and toxic heavy metals. The anionic nature of capsules also attracts nutrients from the surrounding environment. Finally, capsules allow bacteria to adhere to the cells of hosts and various substrates in the natural environment, permitting colonies to grow.

The apparent size of the capsule varies widely among bacterial species. In heavily encapsulated forms such as the pneumococci and *Klebsiella* spp., the thickness of the capsule is frequently greater than the diameter of the bacterial cell. In other bacteria, it may be but a thin layer, closely adherent to the cell's surface.

Bacterial capsules are not stained by usual procedures such as Gram- or acid-fast staining, because the capsule layer fails to retain these dyes.[51] Capsules may be demonstrated under the light microscope using negative-staining techniques, for example, by suspending the bacterial cells in diluted India ink. The capsule displaces the colloidal carbon particles in the ink, causing the cells to appear to lie in lacunae against a dark background (Figure 6).

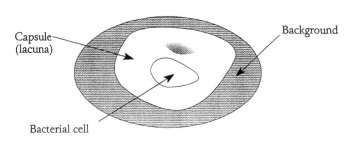

Figure 6

1. Constitution of Capsular Substances

The capsule is, almost always, composed of polysaccharides, varying in both complexity and specific composition, and appearing in both linear and branched chains. Chemically, the simplest consist of homopolysaccharides, which are polymers of a single type of monosaccharide.[47,48] Examples include bacterial celluloses, levans, dextrans, and glucans. Most capsules, however, consist of the more complex heteropolysaccharides, often with uronic acid as an additional component.[44] The monomeric constituents of these polysaccharides may include D-glucose, D-mannose, D-glactose, L-fructose, L-rhammose, ribitol, and glycerol, as well as various amino acids and uroic acid.[48,50] Phosphorous may also be present, particularly in polysaccharides containing polyols (e.g., ribitol and glycerol) and teichoic acids.[52]

Polysaccharides appear in both linear or branched chains. The biosynthesis of most polysaccharides appears to take place in the cell at the cytoplasmic membrane, involving diphosphate sugars as well as isoprenoid lipid intermediates. The polysaccharides are then transported to the extracellular location as capsule material. In various situations, capsule synthesis is less complex. The sugars (e.g., dextran, levans, etc.) are formed by an extracellular process that does not involve the nucleoside diphosphate sugars or the lipid carriers. In other situations, the bacteria is not a polysaccharide but a polypeptide, notably *Bacillis sp.*[47]

The cell wall is found in all bacteria except *Mycoplasma spp.* The cell wall, being rigid, protects the cell from bursting in low osmotic pressure environments and maintains cell shape.[48]

2. Gram-Positive Cell Envelope

Gram-positive bacteria characteristically produce specific surface polysaccharides and proteins in association with a peptidoglycan structural component. The better known polysaccharides include teichoic acids, comprising many of the pneumococcal capsular substances and streptococcal group polysaccharides.[48] Poly-D-glutamic acid polymers are produced by some *Bacillus* species, and the M-protein of Group A streptococci is a virulence factor. Observed under electron microscopy, thin cross-sections of Gram-positive cells reveal a relatively thick, contiguous cell wall layer overlying the plasma membrane. Both proteins and polysaccharides contribute to the layered wall substructure. The Gram-positive cell wall is sensitive to (destroyed by) lysozyme, an enzyme common in skin secretions and tears.

3. Gram-Negative Cell Envelope

Not including a capsule, Gram-negative bacteria exhibit three layers in their cell envelopes (Figure 5). These are the outer membrane, a middle dense layer, and the inner cytoplasmic (plasma) membrane.[35,47] The middle dense layer between the outer and cytoplasmic membranes—termed the piroplasmic space—is occupied by a gel, bounded by a layer of peptidoglycan lying beneath the outer membrane. Both the plasma and outer membranes appear as bileaflet-trilayered sandwich structures, when observed using transmission electron microscopy of membrane cross-sections. A helical lipoprotein, one third of which is covalently linked at one end to the peptidoglycan layer, inserts its lipid end into the outer membrane, thereby anchoring the outer membrane to the cell.[52] The cell envelope can be isolated free of soluble cytoplasm by cell rupture and differential centrifugation. The inner membrane may be dissolved with mild nonionic detergents, leaving the outer membrane bound to insoluble peptidoglycan. The outer membrane can be disrupted by ethylenediamine tetraacetic acid (EDTA), strong ionic detergents, aqueous phenol, or butanol extraction.

4. Outer Membrane

Gram-positive species do not exhibit an outer membrane per se. The outer membrane of Gram-negative species are similar in appearance and contain lipopolysaccharide (LPS), also known as the somatic O surface antigen or endotoxin), phospholipids, and unique proteins that differ from those found in the cytoplasma membrane. Compared to the cytoplasmic membrane, the outer membrane contains less phospholipid, and fewer protein compounds, but contains unique bacterial lipopolysaccharides.[49] The inner and outer leaflets of the outer membrane are also uniquely asymmetrical, appearing in electron micrographs as two electron-dense layers sandwiching an electron neutral (translucent) middle layer. The outer leaflet is functionally distinct for most Gram-negative bacteria. It acts as a hydrophobic diffusion barrier against a variety of substances and it contains receptors for both bacteriophages and bacterocons. The outer leaflet has a participatory role in both cell division and conjunction and it also contains a number of systems mediating the uptake of nutrients and the passive diffusion of small molecules into the piroplasmic spheres. The outer membrane, in conjunction with the thin peptidoglycan layer, provides structural integrity to the bacterial cell. In Gram-negative enteric bacteria (e.g., *Salmonella*), phosphatidylethanolamine occurs almost exclusively in the inner leaflet, whereas the anionic, hydrophilic lipopolysaccharide occurs only in the outer leaflet. In *Neisseria* and *Haemophilus* species, however, phospholipids are found in both the inner and outer leaflets of the outer membrane.[49] Proteins known as porins are distributed throughout the outer membrane and form transmembrane diffusion channels for low molecular weight, water-soluble substances, or serve as bacteriophage (virus) receptors.

5. Lipopolysaccharide

Lipopolysaccharides are part of the outer membrane and are unique to Gram-negative bacteria.[54] They comprise the major surface antigenic components of the somatic O antigens and are responsible for the endotoxin properties of Gram-negative bacterial cells, in general.[47]

Lipopolysaccharides are high molecular weight complexes usually consisting of three regions: a lipid region, termed Lipid A; a core polysaccharide region; and a specific polysaccharide region, the O-specific chains (Figure 7).[47,54]

The O-specific chains project from the surface of the outer membrane and serve to protect the cell from antibody and complement by interfering with their direct contact with the outer membrane.[35] The Lipid A region is responsible for the endotoxin activity of bacterial lipopolysaccharides and is situated in the hydrophobic zone of the outer membrane.

6. Outer Membrane Proteins

Many of the properties of the outer membrane previously described are attributable to the proteins found in the membrane. Those proteins present in the greatest amounts are designated the "major" outer membrane proteins and are generally divided into three groups: (1) the pore forming proteins (porins); (2) the nonporin proteins; and (3) the lipoproteins.[55]

The proteins of the Gram-negative outer membrane act to facilitate the permeation of solutes through the membrane.[28] The simplest, the matrix porins, form transmembrane channels through which hydrophilic solutes of low molecular weight flow. Solutes do not interact with the porins in this passive permeation. Other porins have more specific permeation functions and, in some instances, interact with the substrate as a part of its translocation through the membrane.

Nonporin proteins are not as well understood as porins. Some are known to span the outer membrane and associate with the peptideoglycan layer. These proteins may serve as receptors for a variety of ligands, regulate exopolysaccharide biosynthesis or help stabilize the cell envelope.[55]

The lipoproteins are the smallest of the outer membrane proteins, but are the most common. Lipoproteins are usually covalently bound to peptidoglycan and are essential to structural stabilization of the cell envelope.[35] The role of those lipoproteins shown to be free and unbound is not clearly understood.

Lipid A	Core Region	O-Specific Side Chains

Figure 7 Basic lipopolysaccharide structure of Gram-negative bacteria.

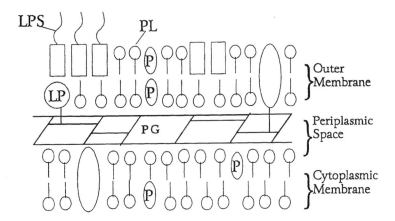

Figure 8 Gram-negative cell envelope. LPS: Lipopolysaccharide; LP: Lipoprotein; P: Protein; PG: Peptidoglycan; PL: Phospholipid.

7. Gram-Positive Cell Wall Structure

Electron microscopy of thin sections of Gram-positive bacteria demonstrate an amphorous, electron-dense layer lacking fine structure throughout. In Gram-positive bacteria, the cell wall is 15–50 mm in thickness, making up 20–40% of the dry weight of the cell. The cell wall of Gram-positive bacteria is composed, to a large degree, of peptidoglycan which contains 20–50 peptidoglycan layers.[48,55]

Within the wall, occurring on its surface layers, are a variety of polysaccharides, proteins, and teichoic acids. Many of these surface components are immunologically reactive, particularly the surface proteins (C-polysaccharides) found on *Streptococci* spp. The teichoic acids of this group are of special concern. They are constituents of the cell wall and membrane having similarities. The basic structure of teichoic is that of a glycerol polymer or ribitol units joined together by phosphodiester linkages (Figure 9).

D-alanine and carbohydrate components may be ether linked to this linear backbone structure. These teichoic acids occur in Gram-positive bacteria as membranes associated (glycerol type) or wall associated (glycerol or ribitol types).[48,50,55]

The membrane associated teichoic acids are covalently linked to glycolipids contained in the plasma (cytoplasmic) membrane. These compounds are commonly termed "lipoteichoic acids." The glycolipid portion of the molecule is attached to the outer leaflet of the membrane with hydrophilic polyglycerol phosphate chains extending outward toward the cell wall.[49]

The wall associated teichoic acids are covalently linked to lycan chains of

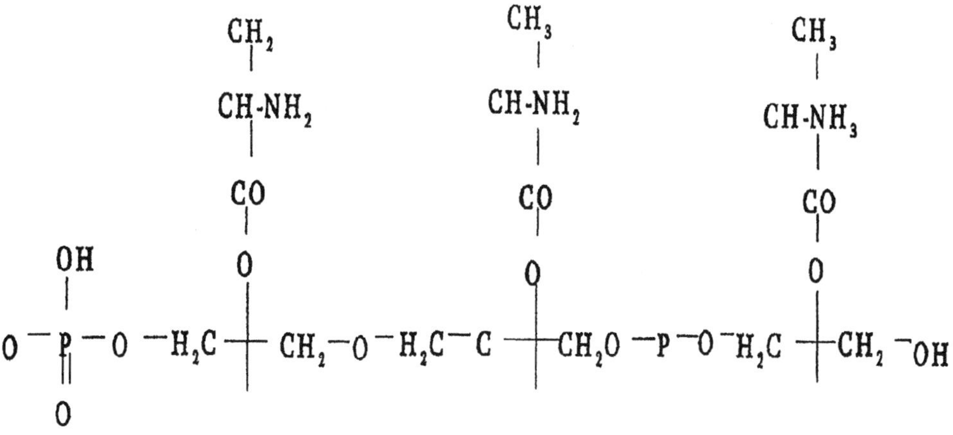

$$CH_2$$

$$CH-NH_2$$

$$CO$$

$$CH_3$$

$$CH-NH_2$$

$$CO$$

$$CH_3$$

$$CH-NH_3$$

$$CO$$

OH

$$O^- \underset{\underset{O}{\parallel}}{P} {}^-O^-H_2C \; + \; CH_2^-O^-H_2C^-C \; + \; CH_2O^-P^-O^-H_2C \; + \; CH_2^-OH$$

Figure 9   General teichoic acid structure.

peptidoglycan contained in the cell wall structure.[53,55] Since these teichoic acids can participate in immunological reactions, they likely are partially exposed on the cell surface. The teichoic acids probably serve to stabilize the cell wall and maintain its contact with the cytoplasmic membrane. Teichoic acid is also thought to contribute to magnesium binding and helps maintain proper ionic range condition for cationic-dependent enzymes in the cell envelope. Since they behave as cationic-binding polymers, they can act as ion exchangers in a manner similar to those found in relation to polysaccharides in the slime layer or capsule production. Their presence on the bacterial cell surface may also enhance mucosal attachment of pathogenic Gram-positive bacteria.[54]

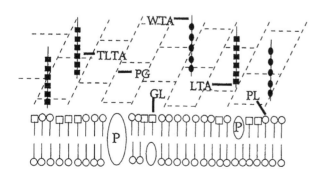

Figure 10   Gram-positive cell wall. LTA: Lipotrichoic acid; TLTA: Transient Lipoteicheic acid; WTA: Wall-Associated Teichoic Acid; P: Protein; PG: Peptidoglycan; PL: Phospholipid; GL: Glycolipid.

While they are not usually found in isolated states, teichoic acids have been shown to induce specific antibody formation when combined with other cell constituents in the complete cell.[56] They are, therefore, important as surface antigens in many Gram-positive bacteria.

The topography of the outer surface of bacterial cells is of considerable interest because they provide insight into localization of specific cell wall components. In this respect, thin section electron microscopy has not been particularly informative but freeze-etching and negative staining has revealed a regularly occurring array of macromolecules on the cell surface designated the S-layer.[51] Most of these surface patterns in Gram-positive bacteria consist of subunit arrays of tetragonal symmetry, with center-to-center spacing ranging between 7–16 nm. These appear to be composed of single subunit types, usually protein in motive covering the entire surface of the cell.

8. Adhesion Sites (Bayer Junctions)

In Gram-negative cells, points of connection between inner and outer membranes are known as adhesion sites, or Bayer junctions.[55] The Bayer junctions are physiologically active.[48] On the outside, they are sites of bacteriophage attachment-DNA injection and complement-mediated lysis. Internally, adhesion sites appear to be growth zones, and they serve as sites for translocation of secretory protein, outer membrane proteins, lipopolysaccharides, and capsular polysaccharides, and as emergence sites for pili and flagella.

9. Periplasm

Periplasm, which occurs in the space between the cytoplasmic and the outer membranes, may readily be observed in Gram-negative bacteria, but not at all or only with difficulty in Gram-positive bacteria.[47,55] This may be explained by the relatively high internal osmotic pressures of Gram-positive bacteria, compared to those of Gram-negative bacteria. The piroplasmic gel space of Gram-negative bacteria varies with growth conditions and among individual bacteria. The gel is quite viscous and possibly highly ordered in structure, surrounds and is interspersed with the porous peptidoglycan. It contains cytoplasmic membrane-derived oligosaccharides that occur in inverse proportion to the osmolarity of the growth medium, various hydrolytic enzymes such as phosphatases, nucleuses, plasmid-controlled β-lactamases (penicillinases), and proteins that specifically bind sugars and transport amino acids and inorganic ions. Release of these constituents from the periplasmic space can be induced by osmotic shock—i.e. rapid dilution of hypertonic cell suspensions—after EDTA treatment.[54]

10. The Cytoplasmic (Plasma) Membrane

Beneath the rigid cell wall layer, and in close association with it, is the delicate cytoplasmic membrane, vitally important to the cell. In thin sections, the plasma membrane demonstrates by electron microscopy a typical trilaminar sandwich structure of dark-light-dark layers.[35]

The bacterial cytoplasmic membrane is similar in chemical composition and structures to that of eucaryotic cells. They are both composed of a phospholipid bilayer with proteins interspersed in the membrane. The membrane is composed of 50–75% protein, 20–35% lipid and, in bacterial cells, makes up about 10% of the dry weight.[48]

The bilayer membrane, with a medium hydrophobic zone, is traversed by proteins that are thought to be permeases involved in the active transport of small substrates (e.g., various amino acids, carbohydrates, etc.) to the cell's interior regions. Cytoplasmic membrane proteins are also associated with both inner and outer leaflets of the membrane. Other proteins of the membrane are related to enzymatic functions, for example, oxidative phosphorylation, macromolecular synthesis of the cell wall, and other less well-known tasks.[53,55] Within the cytoplasmic (inner leaflet) side of the membrane are situated enzymes responsible for various cytoplasmic maintenance functions.

The cytoplasmic membrane lacks the strength and rigidity of the bacterial cell wall. The membrane is incapable of maintaining the cell shape, other than that of a spherical "coccus," if the cell wall is removed.[53] The membrane normally is found just under the cell wall and is held firmly there by the internal osmotic pressure. While the cytoplasmic membrane is structurally distinct from the cell wall, the two are bonded.

11.   Cytoplasmic Membrane as Osmotic Barrier

Although bacteria are regarded as extremely tolerant of osmotic changes in their external environment, their protoplasts (Gram-positive bacteria lacking cell walls) or sphaeroplasts (Gram-negative bacteria lacking cell walls) undergo either plasmolysis (shrinkage) or plasmoptysis (swelling) when placed in appropriate media. Cell-wall-free bacteria are very susceptible to the vagaries of the environment. Placing intact bateria into hypertonic solutions results in plasmolysis, that is, shrinkage of the membrane and cytoplasm from the cell wall. Gramnegative cells are more easily plasmolyzed than are Gram-positive cells, which correlates with their relative internal osmotic pressures.

The osmotic barrier is demonstrated by a bacterium's ability to concentrate certain amino acids against concentration gradients which, in Gram-positive bacteria, may be as much as 300- to 400-fold.[55] Phosphate esters, amino acids, and other solutes contribute to the internal osmotic pressure. Osmotic activity is also indicated by selective permeability toward various compounds.

E.   Membrane Components

Membranes account for some 30% or more of the cell weight.[35,47] Membranes contain 60% to 70% protein, 30% to 40% lipids, and small amounts of carbohydrate. Phosphatidylethanolamines (75%), phosphatidylglycerol (20%), and glycolipids are found as major constituents. Choline, sphingolipids, polyunsaturated fatty acids, inositides, and steroids are generally absent.[56] Pathogenic *Mycoplasma spp.*, however, do incorporate steroids from the environment into their plasma membranes. Glycolipids, including diglycosydilglycerides, are found primarily in Gram-positive bacterial membranes, which also contain lipoteichoic acids.

Various enzymic activities are associated with membrane proteins. These include the energy-producing bacterial cytochrome and oxidative phosphorylation systems, membrane permeability systems, and various polymer-synthesizing systems. An ATPase has been isolated from knob-like membrane structures similar to those found in eukaryotic mitochondria.[55,56] Up to 90% of the cellular ribosomes may be isolated as a membrane-polyribosome-DNA aggregate.

1. Mesosomes

The cytoplasmic membrane may be invaginated to form internal organelles known as mesosomes.[55] These can assume a vesicular, lamellar, or tubular form, and more than one of these forms may be observed together in a bacterial cell. The gross chemical composition of mesosomes is equivalent to the cytoplasmic membrane. They occur in all three forms in Gram-positive bacteria but the laminar type is typically observed in Gram-negative bacteria. Although many functions have been proposed for mesosomes, none have been firmly established. However, mesosomes have been linked largely on morphological grounds, with cell replication and apportionment of deoxcyribonucleic acid (DNA) to daughter cells in cell division. Mesosomes also act as septum initiators during cell division.

2. Peptidoglycan Layer

The shape and rigidity of bacterial cells are due almost entirely to the presence of a lare polymeric supporting structure called peptidoglycan (also murein, mucopeptide, or glycosaminopeptide) that lies just outside to the cytoplasmic membrane.[48,55] There are only a few species of bacteria which lack this structure but, in particular, the *Mycoplasma* spp., which lack cell walls in situ.

The peptidoglycan layer can be thought of as a single sack-like, macromolecular structure, completely encompassing the cytoplasmic body of the cell. The peptidoglycan layer is composed of glycan strands with short peptide linkages, and the strands themselves are composed of alternating units of β-1, 4-linked N-acetyl-D-glucosamine-and N-acetyl-D-muramic acid, the 3-lactylether of D-glucosamine.[55,56] The chain lengths of individual glycan strands vary from <10 to >170 disaccharide units (Figure 11).

The glycan strands are cross-linked by short peptide chains.[48,55] The N-terminus of the peptide subunit is bound through the carboxyl group of muramic

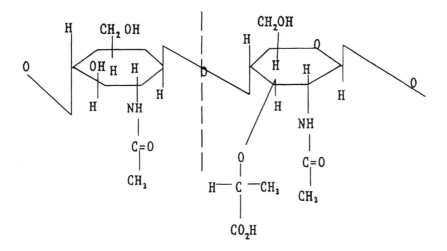

N-acetylglucosamine   N-acetylmuranic acid

Figure 11   Disaccharide structure of peptidoglycan.

acid and the amino acid sequence of the linking peptide is usually L-alanine-to D-isoglutamine to meso-diaminopimelic acid (or L-lysine) to D-alanine. These peptide subunits of adjacent glycans are directly joined through the unbound amino group of diamino pimelic acid and the C-terminal of D-alanine or indirectly through an interpeptide bridge. Although there is a remarkable degree of constancy in the glycan strands, different groups of bacteria show variation in the composition of the interpeptide bridges and, to a lesser degree, in the composition of the peptide chains.[48]

The peptidoglycans are susceptible to the hydrolytic activities of a number of enzymes, many of which are produced by potential hosts for pathogenic bacteria. The most studied of these is lysozyme (N-acetylmuramidase), which hydrolyzes the glycan strand at the glycosidic linkages between N-acetylmuramic acid and N-acetylucosamine (Figure 11). Lysozyme and other similar enzymes thereby depolymerize the peptidoglycan and, in Gram-positive bacteria in particular, destroy the cell wall, producing cell lysis.[48,55]

F.   Bacterial Endospores

The endospore is an oval to spherical body formed within bacilliform bacteria that represents a dormant state highly resistant to lethal effects of heat, drying, cold, lack of nutrients, and many chemicals lethal to the vegetative forms of the bacteria.[50,52] Its high resistance to lethality provides the bacteria with a survival advantage over nonspore-forming species of bacteria.[57] Endospore formation, however, is uncommon in bacteria and is limited to the species of *Bacillus* and *Clostridium*, plus a few minor genera of aerobic and anaerobic bacteria. The *Bacillus* spp. and *Clostridium* spp. are widely distributed in nature. Endospores are light-refractile, and their size, shape, and position in the mother cell are relatively constant characteristics of a given species.[49] Endospore coats include a rigid peptidoglycan layer, which differs in composition from that of the parent vegetative cell in that peptide cross-linkages are markedly reduced and large amounts of dipicolinic acid and calcium ions are incorporated.[57] Surface antigens of endospores usually differ from those of the vegetative cell.[35,47] Additionally, because endospore formation represents a form of cell differentiation, it is of interest to cell biologists, as well as to specialists in sterilization technology who use the resistant spores as indicators of sterility.[57]

Within the vegetative cell, the endospore is spheroidal or oval with the long axis parallel to that of the bacillus. Its breadth may be the same as the vegetative cell, or even greater, providing a bulging appearance to the cell as commonly seen in *Clostridium* spp. The relative size of the endospore is constant, as its position in the vegetative cell. Endospores may be located in the center of the vegetative cell (central), part-way between the center and the end of the cell (subterminal) or at the end of the cell (terminal).[55,57] Hence, endospore size and location result in the characteristic morphology of the microorganism. For example, *Clostridium tetanal* have a "drum-stick" appearance because of the large terminal spore.

Endospores can be observed in the living bacteria by phase contrast light microscopy. In fixed (killed) preparations, endospores do not stain using the Gram-stain procedure, but require heating for the stain (usually carbol-fuchsin) to penetrate them.[51] Endospore structure is complex.[57] The protoplast (spore core), the innermost portion, is bound by a limiting membrane analogous to the cytoplasmic membrane of the vegetative cell. A cortical layer (cortex) encloses the protoplast. The cortex, in turn, is enclosed by one or more complex, cystine-rich protein layers termed the spore coat(s), that are unique to spores, containing compounds not found in vegetative cells. Finally, the endospore may be surrounded by a thin, often loose-fitting bag-like structure, the exosporium. Figure 12 provides a diagrammatical representation of an endospore.

1. Dormancy

Bacterial endospores are physiologically inactive.[57] In many cases, there is no significant endogenous metabolism but, in some, minimal metabolic reactions are detectable. Even so, it is clear that the endospore has the required metabolic capacity for both germination and vegetative growth. Dormancy may be associ-

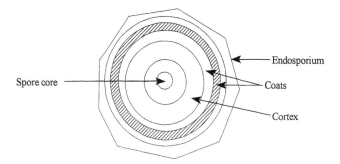

Spore core

Endosporium

Coats

Cortex

Figure 12 Diagram of a typical endospore.

ated with the presence of large amounts of the unique chemical compound, dipicolinic acid in the protoplast. Chelated with calcium ions, dipicolinic acid is also important in the structure of the endospore cortex and intercalated within the helical structure of DNA, it displaces intramolecular water. The heat resistance of bacterial spores is thought to be due to the reduced amounts of water in the densely-configured, greatly compressed core structure. Notably, dipicolinic acid is virtually absent from the vegetative cells prior to sporagenesis.

2. Bacterial Nucleus

Using the Feulgen strain, bacterial DNA can be detected as nucleoids or chromatin bodies by light microscopy.[35,47] It is difficult to demonstrate chromatin bodies by direct staining because of the high concentration of RNA in the bacterial cell, but this can be removed by using ribonucleases. Chromatin bodies have been shown to be present at all stages of bacterial growth.

Electron microscopy of thin sections of bacteria reveal that the nuclear material is an irregular, thick, fibrillar DNA network. Often running parallel to the axis of the cell, a direct attachment of the cell membrane can be clearly observed. During multiplication, bacterial DNA remains as a diffuse chromatin network and never aggregates to form a well-defined chromosome as is observed in the mitosis of eucaryotic cells.

3. Ribosomes

Observation of negatively-stained thin sections of bacterial cells under electron microscopy reveals the presence of 70 Svedberg-unit (70S) ribosomes, each of which is comprised of a 30S and 50S subunit.[55,56] The ribosomes are composed of ribosomal RNA (rRNA). The 30S subunit is responsible primarily for decoding the messenger RNA (mRNA) for protein synthesis. The 50S subunit contains the 23S and 5S rRNA molecules which are involved primiarily with peptide bond formation in the assembly of amino acids into peptide chains.[48] The ribsomes tend to form aggregates of varying sizes called polysomes, that are often bound with strands of mRNA. Ribosome numbers vary greatly according to environmental growth conditions.

III. MYCOLOGY (FUNGI)

Fungi, unlike bacteria, are eucaryotic cells that contain at least one nucleus, nuclear membrane, endoplasmic reticulum and mitochondria.[47,58] Fungal cells are much larger than bacteria and are closer structurally and metabolically to cells of higher plants and animals.[59] Most fungi possess a rigid cell wall and some species produce motile, flagellated cells. Unlike most members of the plant kingdom, fungi are nonphotosynthetic.[60]

Fungi may exist as single oval cells (yeasts), that reproduce by budding, or of long tubular strands (hyphae), usually septate, which exhibit apical growth and true laterial branching.[35,47,58] Branching intermingled and often fused, overlapping hyphae constitute the mycelium which forms the visible mold colony.[59,60] Reproduction may be vegetative (asexual) via variously specialized germinal cells called conidia, or it may consist of elaborate specialized structures which facilitate fertilization, protection and dissemination of the resultant spore. Conidia may be simple fragments of a hypha, or they may be produced from variously specialized structures called coniophores. Spore-bearing structures may be simple lateral branches of the hyphae or they may be constructed into large reproductive bodies such as mushrooms and bracket fungi which protect the spores as they develop and facilitate their dispersal at maturity.

The natural habitats of most fungi are water, soil, and decaying organic debris, and most are obligate or facultative aerobes.[59,61] They are chemotropic organisms, many of which secrete enzymes which degrade a wide variety of organic substrates into soluable nutrients that can be absorbed.

A. Yeasts

Yeasts are single cells, usually spherical to ellipsoidal ranging from 3 to 15 μm in diameter.[35,58] Most reproduce by budding, but a few do reproduce by binary fission.

After growth on agar media (24–72 hours), yeasts tend to produce colonies which are pasty and opaque-looking, generally growing in colonies from 0.5 to 3.0 mm in diameter. A few species have characteristic pigments, but most are cream-colored. Using microscopic and colonial morphology, it is difficult to distinguish species, so nutritional studies and specialized media must be used.[59,60]

B. Molds

The molds are growth forms which are multicellular and filamentous and appear in colonies.[59,60] The colonies consist of branching cylindrical tubules varying in diameter (2 to 10 μm) termed hyphae, which grow by apical elongation. Hyphal tips contain densely packed membrane-enclosed vesicles, many of which fuse with the cell membrane during active growth.[47] The mass of intertwined hyphae that accumulates and consitutes the visible mold is termed the mycelium. The hyphae of some species are divided into cells by cross-walling (septa) that form at regular, repeating intervals during filamentous growth and, in some, the septa are penetrated by pores that permit flowing of cytoplasm but not organelles.

Molds are extreme opportunists, tending to grow well on a tremendous variety of substrates found in the environment and on most laboratory culture media.[59,61] Hyphae which actually penetrate into the substrate/media are termed vegetative of substratic hyphae and serve to anchor the mycelium. Those hyphae which extend above the surface (aerial hyphae) usually bear the specialized structures (conidiophores) that produce the asexual reproductive cells, the conidia.

C. Dimorphism

Most fungi exist only as a mold or a yeast form, yet a number of species, including several important pathogens in humans, are capable of growing as both a yeast and a mold form.[35,47,58] Temperature plays a major role in this dimorphism. At about 35–37°C, such species grow in the yeast phase but, at lower temperatures (20–30°C), they grow as molds. Although this is referred to as thermal dimorphism, available nutrients, carbon dioxide levels, cell density, age of culture, or a combination of these factors can also induce a shift in morph.

D. Cell Structure

Fungal cells are composed of a cell wall, cell membrane and cytoplasm containing an endoplasma reticulum, nuclei, nucleoid, storage vacuoles, mitochondria, and other organelles (Figure 13).

E. Capsule

Some fungi secrete an external coating of slime.[58,61] This slime layer, or capsule, is composed mostly of amphorous polysaccharides that play a major role in the cell's clumping. The capsular composition varies among species in quantity, chemical composition, antigenic properties, and physical attributes such as solubility and viscosity. The capsule component does not affect the permeability of

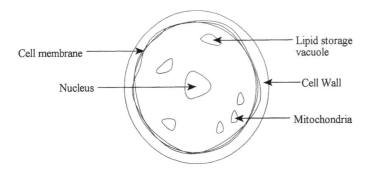

Figure 13 Basic fungal structure.

the cell wall and cell membrane but, due to its slime nature, it probably influences growth by preventing the disassociation of buds from yeast cells and the dispersion of yeast in air and water. As in bacteria, the capsule is thought to be associated with virulence of those fungi infectious to man or animals.

F. Cell Wall

The cell wall is 15–30% of the dry weight of a fungal cell.[35,47,49] The cell wall provides rigidity and strength to the cell as well as prevent it from osmotic shock. As in bacteria, the cell wall determines the morphological shape of the cell. The cell wall is, on the average, thicker in yeasts than in the hyphae molds.

About 80% of the cell wall is composed of carbohydrate material.[47,58] Several different types of polysaccharides are found and these vary in type and relative abundance from species to species. The major polysaccharide components in fungal cells include chitin, chitosan, cellulose, β-glucan, α-glucan, and mannan.[35,47,55] Since the polysaccharides shared by fungi are similar, many fungi groups exhibit the same surface antigens. However, the side chains of the glucans vary considerably in the number, length and linkage of their residues, and therefore, present many unique antigens. Hence, there are many common antigens unique to species, as well as those shared in common with other species.

About 10% of the fungal cell wall is composed of protein and glycoprotein.[35,55] Proteins include enzymes involved in cell wall growth, specific extracellular enzymes and structural proteins which cross-link the polysaccharide chains. The concentration of protein increases toward the inner cell wall surface. Cell wall proteins are high in sulfur-containing amino acids linked by disulphide bonds, and these are more common in walls of hyphae than in yeast walls. The reduction in disulfide bonds appears to play a role in the transformation from mold stage to yeast stage of growth, among the thermally dimorphic species.

Cell wall polysaccharides are fibrillar and multilayered, appearing as long microfibrils microscopically.[59,60] Four to eight distinct layers are usually observed in cells walls, but the degree of organization is variable. Generally, the most compact, most protein-laden layer is nearest to the cell membrane.[35,47,59] The external layers tend to be more amphorous, less well-organized, and less compact. Some fungal cell walls contain tightly interwoven microfibrils embedded in an amphorous, polysaccharide matrix.

In infection, the host is initially exposed to the fungal cell surface and research suggests that the cell wall governs much of the pathogenicity of the fungus.[58,61] The cell walls of the major pathogenic fungi have been shown to possess components that mediate host cell attachment (e.g., phagocytes, epithelial cells, and endothelial cells). Cell surface ligands, receptors, and other cellular components promote colonization and invasion of host cells and resistance to host immunological defenses.[62]

Host immune response to fungal cell wall antigens is very strong and quite specific for a number of cell wall determinants, some of which are specific oligosaccharides.[35,47,62] Mammalian tissues lack the enzymes necessary to degrade many cell polysaccharides and, therefore, fungal cells can be eliminated from the body only very slowly.

G. Cytoplasmic Membrane

Fungi have a bilayered, cytoplasmic membrane similar to that of higher eukaryote cells.[55,58] The membrane protects the cytoplasm, regulates the intake and secretions of solutes, and assists in the synthesis of the cell wall and capsule components. The membrane contains several phospholipid types, their relative proportions depending upon the species of fungus. The most common include phosphatidycholine and phosphatidylethanolamine, but phosphatidylserine, phosphatidylinositol, and phosphatidylglycerolare also found in lesser quantities.[55] The total phospholipid content of the cytoplasmic membrane varies between species, between strains of a species, and even within those strains, depending on environmental conditions.

Fungal cytoplasmic membranes contain sterols, in contradistinction to those of bacteria (except *Mycoplasma* spp., which do not actually produce the sterols, but acquire them from host cells). The principle sterols found in fungi are ergosterol and zymosterol, as opposed to cholesterol, the principle sterol of mammalian cells.

H. Cytoplasmic Content

Often, multiple nuclei are seen in both yeast and mold cells.[48,50] For example, when septa are not present in mold hyphae (the zygo mycetes), there is a plasmic continuity, and the hyphae are plainly considered multinuclear. Among the "higher" fungi, the septate hyphae often possess a central pore in the septum that allows "mixing" of cytoplasmic content. Many species are capable of dilation/ constriction of the septal pores, allowing the flow of organelles, including nuclei, mitochondria, lipid vacuoles, and ribosomes, between the hyphal cells.

Fungal mitochondria resemble those of both plant and animal eukaryotic cells. The number of mitochondria within each cell can vary and is directly related to the level of cellular respiration. During sporulation, for example, which requires increased energy, the number of mitochondria increase within each cell. In tissue, fungi tend to have fewer mitochondria.[60]

The cells of many fungal species contain very characteristic vacuoles that function as complex organelles. Some vacuoles contain a variety of hydrolytic enzymes, while others serve as storage depots for ions and metabolites such as amino acids, polyphosphates, and other compounds.

IV. VIRUSES

Viruses are a unique life form quite distinct from the procaryotic or eucaryotic organisms discussed earlier.[35,47] They are smaller than bacteria and fungi and are obligate parasites within the cells of other organisms. They consist of a single type of nucleic acid—DNA or RNA, either single- or double-stranded—encased by a shell (capsid) variously structured and composed primarily of protein. The capsid of some viruses is further enclosed by a unit membrane composed of lipids and glycoproteins (so-called "spike proteins").[35,53,55] Viruses do not reproduce by binary fission because they lack the intracellular components necessary to produce macromolecules. Instead, viruses "capture" and redirect the organelles of a host cell to synthesize new virus components, which then autoassemble within the host cell into virions, the complete viral particle.[35,47,51,52]

Virions range in size from 20–25 mm. Viral nucleic acid is encased by a capsid that generally takes the form of either a helix or an icosahedron, although some viruses take neither form, and are termed complex viruses. The nucleic acid/capsid is referred to as the nucleocapsid. The capsid itself is composed of single polypeptide chains, termed structural units, which may aggregate to form the multiple polypeptide units termed capsomeres, or combine directly to form the helix-form capsids.[52,62] The capsomere combine to form the more complex icosahedral capsids. The virion, a mature infectious virus particle, may consist of the nucleocapsid (capsid containing nucleic acid) alone, or may be surrounded by an envelope of lipids and glycoproteins acquired during the process of "budding" through the cytoplasmic membrane of the host cell.[50,62]

Viruses are generally grouped into five morphological categories.

Naked Icosahedral Viruses

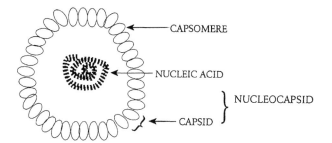

Enveloped Icosahedral Viruses

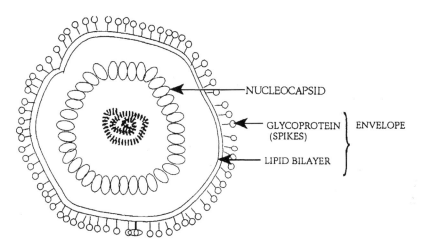

Naked Helical

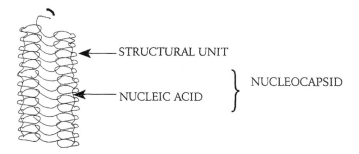

Enveloped Helical

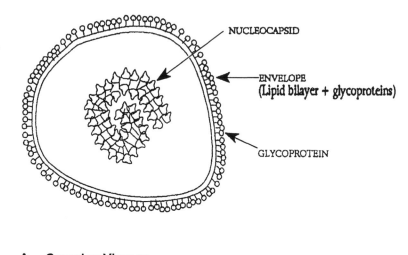

A. Complex Viruses

The fifth category incudes a number of viruses that do not fit into any of the previous four groups and are termed complex viruses. The poxviruses are examples of the complex structural form.[53]

4

Current Topical Antimicrobials

In this chapter, we discuss various aspects of the topical antimicrobial products currently in common use in medical, food service, and consumer (personal hygiene) markets. The antimicrobial products of primary interest include iodine complexes (aqueous iodophors and tinctures), aqueous formulations of chlorhexidine gluconate, triclosan, and parachlorometaxylenol (PCMX) alcohol formulations and tinctures of chlorhexidine gluconate.[2,33]

I. IODINE COMPLEXES

Iodine in its pure form is relatively insoluble in water without a solubilizing agent, but it dissolves well in various alcohols to provide an iodine tincture.[64] Tinctures of iodine are used primarily as antiseptics.

By far, the most common form of iodine for use as a topical antimicrobial is the iodophor.[65] Iodophors are complexes of elemental iodine (tri-iodine) linked to a carrier having several functions: (1) increased degrees of solubility in aqueous solution for the iodine; (2) provision of a sustained release reservoir of the iodine; and (3) reduced equilibrium concentrations of free iodine. The most commonly used carriers are neutral polymers, polyacrylic acids, polyether glycols, polyamides, polysaccharides, and polyalkalines.

The most commonly used iodophor is povidone iodine, a compound of 1-vinyl-2-pyrrolidinone polymer with available iodine ranging between 9% and 12% (United States Pharmacopeia XXIII).[66] The chemistry of aqueous solutions of iodophors is complex. Based on our current understanding, various electronic and steric effects are responsible for the interaction observed between polymeric organic carrier molecules and iodine (Equations 1–5).[67]

$$I_2 + H_2 \rightleftharpoons OI^+ + I^- \text{ (hydrolytic ionization)} \tag{1}$$

$$H_2 + OI^+ \rightleftharpoons HOI + H^+ \text{ (dissociation of } H^+) \tag{2}$$

$$HOI \rightleftharpoons OI^- + H^+ \text{ (dissociation of } HOI) \tag{3}$$

$$3HOI \rightleftharpoons IO_3^- - 2I^- 3H^+ \tag{4}$$

$$I_2 + I^- \rightleftharpoons I_3 \text{ (formation of triiodide)} \tag{5}$$

This interaction permits high degrees of variation in the equilibrium concentrations of iodine. Hence, it is assumed that in interaction with low molecular oxygen compounds such as amides, esters, ketones, or ethers, donor-acceptor complexes are formed with the iodine playing the part of the acceptor.

$$\zeta = O + I_2 \rightleftharpoons \zeta = O - I - I \tag{6}$$

The relatively low concentration of free molecular iodine in povidone iodine concentrates is a reason that it may not be capable of autosterilization.[67] For example, *Pseudomonas* spp. and *Burkholderia cepacia* have reportedly been isolated from iodophor concentrates. Apparently, the bacteria are protected by biofilms.[68]

The amount of free molecular iodine is highly relevant to the antimicrobial effectiveness of iodophors. The differing composition of pharmaceutical additives (e.g., detergents and emollients) which have iodine complexing properties, as well as the ratio of free iodine to total iodine in various proprietary formulations can result in great differences in the concentrations of free molecular iodine available for binding. Hence, the antimicrobial index of importance for a product is the "total available iodine."[64]

Iodine preparations play a variety of roles as skin disinfectants.[65] For example, iodophors are commonly used as preoperative skin preparations, as surgical scrub solutions, and as preinjection preparations, as well as preinsertion preparations for venous/arterial catheters. Iodophors also have been used successfully as therapeutic agents in treating wound infections, including those encountered in burn patients, and for disinfection of medical equipment, such as sutures, catheters, scalpels, plastics, rubber goods, brushes and thermometers. Finally, iodophors have been used successfully as an antimicrobial treatment for drinking water, swimming pool water, and waste water.

A. Range of Action

Iodophors and tinctures of iodine provide excellent immediate antimicrobial action against a broad range of viruses, both Gram-positive and Gram-negative bacteria, fungi, and various protozoa.[67] In fact, almost all important human disease microorganisms, including enteric bacteria, enteric viruses, protozoan trophozoites and cysts, mycobacteria, spores of *Bacillus* spp. and *Clostridium* spp., and many fungal species are susceptible to free iodine. It should be noted, however, that exposure times and concentrations of available iodine required vary (Table 1).[67]

In topical application to skin surfaces (e.g., hands and body surfaces in

Table 1 Some Recommended Applications for Iodine-based Antimicrobials

Scope of application	Concentration	Conditions	Exposure time (min.)	Disinfective result	References
General germicidal action	1:20,000	Absence of organic matter	1	"Most bacteria are killed"	Goodman and Gilman, 1980
	1:20,000	Absence of organic matter	15	"Wet spores are killed"	Goodman and Gilman, 1980
	1:200,000	Absence of organic matter	15	"Will destroy all vegetative forms of bacteria"	Goodman and Gilman, 1980
Disinfection of skin	1% tincture	—	90 sec.	"Will kill 90% of the bacteria"	Goodman and Gilman, 1980
	5% tincture	—	60 sec.	"Will kill 90% of the bacteria"	Goodman and Gilman, 1980
	7% tincture	—	15 sec.	"Will kill 90% of the bacteria"	Goodman and Gilman, 1980
	1% aqueous I_2-solution	Skin of hands	20 min.	"Inactivation of rhinovirus"	Carter et al., 1980
	2% aqueous I_2-solution	Skin of hands	3 min	"Inactivation of rhinovirus"	Carter et al., 1980

the inguinal, abdominal, anterior cubital, subclavian, and femoral regions), iodo-phors and tinctures of iodine providing at least 1% available iodine demonstrate effective immediate and persistent antimicrobial properties.[67] However, it should be noted that in general, neither provides residual antimicrobial action.[15,32]

II. CHLORHEXIDINE GLUCONATE

Chlorhexidine gluconate (CHG) was first synthesized in 1950 by ICI Pharmaceutical in England.[70] CHG was found to have high levels of antimicrobial activity, but relatively low levels of toxicity to mammalian cells.[69,70] Additionally, CHG has a strong affinity for skin and mucous membranes. As a result, CHG has been used as a topical antimicrobial for wounds, skin-prepping, and mucous membranes (especially in dentistry), where it provides, by virtue of its proclivity for binding to the tissues, extended antimicrobial properties. CHG also has value as a product preservative, including ophthalmic solutions, and as a disinfectant of medical instruments and hard surfaces.

CHG is a cationic molecule which is generally compatible with other cationic molecules such as the quaternary ammonium compounds.[70] Some nonionic substances such as detergents, although not directly incompatible with CHG, may inactivate the antimicrobial properties of CHG, depending upon the compound and concentration levels. CHG is "incompatible" with inorganic anions, except in very dilute concentrations, and may also be incompatible with organic anions present in soaps containing sodium lauryl sulphate, and with a number of pharmaceutical dyes.[64,65,70]

The antimicrobial activity of CHG is pH-dependent, with an optimal use range of 5.5–7.0, a nice match with the body's usual range of pH. However, the relationship between antimicrobial effectiveness of CHG and pH varies with the microorganism.[70] For example, CHG's antimicrobial activity against *Staphylococcus aureus* and *Escherichia coli* increases with an increase in pH, but the reverse is true for *Pseudomonas aeruginosa*.

The antimicrobial activity of CHG against vegetative forms of both Gram-positive and Gram-negative bacteria is pronounced.[70,71] It is generally inactive relative to bacterial spores, except when they are exposed at elevated temperatures. Mycobacteria—acid-fast—microorganisms reportedly are inhibited, but not killed by CHG in aqueous solutions. A variety of lipophilic viruses (e.g., herpes virus, HIV, influenza virus) are rapidly inactivated by exposure to CHG. Finally, certain fungi, particularly those in the yeast phase, are sensitive to CHG.

A. Antimicrobial Action

At relatively low concentrations, CHG exerts bacteriostatic effects on bacteria. At higher concentrations, CHG demonstrates rapid bactericidal effects. However, the precise effects vary from species to species and as a function of concentration of CHG.[70,71]

The microbicidal effects occur in a series of steps related to both cytological and physiological changes, which culminate in the death of the cell.[69,70] CHG is known to have an affinity for bacterial cell walls and is absorbed into certain phosphate-containing cell wall compounds. By this process, the CHG is thought to penetrate the bacterial cell wall, even in the presence of cell wall molecular exclusion mechanisms. Once cell wall penetration has occurred, the CHG is attracted to the cytoplasmic membrane. Upon penetration of the cytoplasmic membrane, low molecular weight cellular components (e.g., potassium ions) leak out of the membrane, and membrane-bound enzymes such as adenosyl triphosphatase are inhibited. Finally, the cell's cytoplasm precipitates, forming complexes with phosphated compounds, including ATP and nucleic acids.

As a rule of thumb, bacterial cells carry a total negative surface charge.[69,70] It has been observed that, at sufficient CHG concentrations, the total surface charge rapidly becomes neutral and then positive. The degree of the shift in the charge is directly related to the concentration of CHG, but reaches a steady state equilibrium within about five minutes of exposure. The rapid electrostatic attraction between the cationic CHG molecules and the negatively-charged bacterial cell surface contributes to the rapid reaction rate, that is, the rapid bactericidal effects exerted by CHG.

CHG at antiseptic concentrations (0.5–4%) demonstrates a high degree of antimicrobial effect, both -static and -cidal, on vegetative phases of Gram-positive and Gram-negative bacteria, but it has little sporicidal activity.[70] While there have been concerns that prolonged use of CHG could lead to reduced sensitivity and, ultimately, the development of resistant strains of bacteria, this concern has not been verified, even after prolonged and extensive use.[69,70] There is no evidence that plasmid-mediated antimicrobial resistance, particularly common in Gram-negative bacteria, has developed. This has been borne out in studies of common indicator species such as *Escherichia coli*, *Pseudomonas aeruginosa*, *Serratia marcescens*, and *Proteus mirabilis*. Although several researchers have reported a reduced sensitivity to CHG among certain methicillin-resistant strains (MRS) of *Staphylococcus aureus*, at clinical-use concentrations, this concern has not been substantiated. MRS strains are as susceptible to CHG as are non-MRS strains.

Because viruses have no synthetic properties of their own, the action of CHG is restricted to the nucleic acid core or the viral outer coat. Some viral coats consist of protein and others, lipoprotein or glycoprotein. Outer envelopes that enclose the vital coat of some viruses are mainly of lipoprotein.

In general, CHG is a highly effective antimicrobial in its immediate, persistent and residual properties.[69,70] Higher concentrations of CHG (4% to as low as 0.5%) provide excellent immediate and persistent action, with the added benefit (when used repeatedly over time) of good residual effects. CHG at lower concentrations (less than 0.5%) provide antimicrobial action comparable to PCMX and Triclosan.

CHG, by virtue of its residual antimicrobial properties, may be useful for full or partial bodywashes prior to elective surgery. If the product is used repeatedly over the course of at least 3 to 5 days, the resident microbial populations are reduced by about 3 logs.[15] Hence, when a person undergoes surgery, the remaining microbial populations residing on the proposed surgical site have been significantly reduced, leaving far fewer microorganisms to be eliminated by preoperative prepping procedures.

Currently, there is much interest in alcohol tinctures of CHG.[2,33] These alcohol/CHG products may prove to be highly effective for use as preoperative skin-preps, surgical scrubs, and healthcare personnel handwash formulations. Additionally, tincture of CHG may be useful as both a preinjection and pre-arterial/venous catheterization prep. Preparations of alcohol/CHG combine the excellent immediate antimicrobial properties of alcohol with the persistence properties of CHG to provide a clinical performance superior to either alcohol or CHG alone.

III. PARACHLOROMETAXYLENOL (PCMX)

PCMX is one of the oldest antimicrobial compounds in use, dating back to 1913. It has not been widely used as a surgical or presurgical skin preparation because of its relatively low antimicrobial efficacy.[73]

Because of the absence of substantive data in 1972, the FDA did not designate PCMX as a safe and effective antimicrobial, and it was not formulated at the time for medical purposes such as surgical scrubs, preoperative skin preps, or healthcare personnel handwashes. The initial evaluations from the FDA listed PCMX as a Category III product, meaning that there were not enough data to recognize it as both safe and effective as a topical antimicrobial.[73]

Studies since 1980 have demonstrated PCMX to be safe for human use. After this determination, a number of companies became interested in developing PCMX for use as a topical antimicrobial, and later studies demonstrated PCMX to be an effective antimicrobial compound.

Current over-the-counter formulations demonstrate varying degrees of antimicrobial efficacy, depending upon the formulation. Generally, PCMX products achieve fair to good immediate effects and fair persistent effects, but like iodophors, they provide no residual antimicrobial activity. Currently, PCMX products are mainly formulated for healthcare personnel handwash application. These are effective in removing transient (contaminant) microorganisms from the hands and have low skin irritation properties.

IV. ALCOHOLS

There is much debate concerning the effectiveness of alcohol as a skin antiseptic.[73,74] The antimicrobial efficacy of alcohol is highly dependent on the concentration used, as well as the moisture level of the microbial environment treated. The short chain, monovalent alcohols—ethanol and isopropanol—are probably the most effective for skin disinfection because they are highly miscible with water, have low toxicity and allergenic potential, are fast-acting and are microbicidal, as opposed to microbistatic.

The microbicidal activity of the alcohols is largely a function of their ability to coagulate proteins. The literature suggests, however, that microbicidal effects of alcohols are also a result of their solubility in lipids. Protein coagulation takes place on the cell wall and the cytoplasmic membrane, as well as among the various plasma proteins.

Alcohols are generally inactive against bacterial spores. And, although there is much controversy in the literature concerning the efficacy of alcohols against viruses, there appears to be a general agreement that enveloped, lipophilic viruses are more susceptible to inactivation by alcohols than are "naked" viruses. Lastly, the fungicidal properties of alcohols vary among fungal species, but in general, alcohols demonstrate a relatively high degree of mycidal/-static activity.

Although alcohols, as topical skin disinfectants, provide excellent immediate antimicrobial activity, they have virtually no persistent or residual properties.[73] Hence, their value as surgical scrubs and preoperative skin preps is seriously limited.[2,33] Alcohols have been shown to provide adequate results as healthcare personnel handwashes or preinjection skin preps in removing or killing transient microorganisms, but when used at strengths of 70% and greater, they tend to be drying to the skin, resulting in significant irritation.

V. TRICLOSAN

Triclosan, like PCMX, provides varying degrees of antimicrobial efficacy, depending upon the specific formulation and the species of organism tested.[2,33] Triclosan has been formulated for a wide range of applications and is currently used in healthcare personnel handwash formulations, in the food industry to cleanse workers' hands, and in consumer product lines, including hand soaps, shower gels and body cleansers. Triclosan, like PCMX, provides fair immediate and persistent effects, but no residual action.

5

Measurement of Antimicrobial Action of Topical Antimicrobials

Before going into detailed discussion of evaluative procedures, let us look at the general performance characteristics we intend to evaluate.

I. GENERAL PERFORMANCE CHARACTERISTICS

Generally, topical antimicrobials are evaluated for their antimicrobial efficacy in terms of three parameters. These are: (1) the immediate degerming efficacy of the formulation; (2) the persistent antimicrobial efficacy of the formulation; and (3) the residual antimicrobial efficacy of the formulation, when appropriate.[32,75,76,77]

A product's immediate antimicrobial efficacy is a quantitative measurement of both the mechanical removal of microorganisms by washing and the product's ability to rapidly inactivate microorganisms residing on the skin surface.[32] The persistent antimicrobial efficacy is a quantitative measurement of a product's ability to prevent microbial recolonization of the skin surfaces, either by microbial inhibition or lethality after application of the product. The residual efficacy is a measurement of the product's cumulative antimicrobial properties after it has been used repeatedly over time. That is, as the product is used over time, it is absorbed into the stratum corneum of the skin and, as a result, prevents recolonization of the skin surfaces by microorganisms.

The ability to measure these three parameters accurately poses certain problems to the investigator.[32] Since a product's antimicrobial activity is relative, it is of utmost importance that its efficacy measurements be well-defined and clearly stated before conducting an evaluation.[33] Let us turn our attention to some additional, important points to consider when evaluating topical antimicrobial products.

Traditionally, topical antimicrobial evaluations have been, logically and appropriately, the domain of the microbiologist, since evaluations are made in terms of microbiological parameters.[32] However, although the focus is on microbiology, the efficacy of the product is measured mathematically, specifically through bio-statistical methods. Hence, because of the bipartite structure necessary in these evaluations, the investigator must have a strong background not only in clinical microbiology, but in biostatistics as well, to design a study that will unambiguously answer the experimental questions of interest.

Because it is the experimental design and the statistical evaluation that customarily give investigators the most trouble, we will concentrate mainly in these areas. Our general strategy, then, will be to design, conduct, and evaluate the study by using experimental design concepts, microbiology, and statistics, as appropriate.[2,32]

A statistically designed study is one that systematically collects, organizes, analyzes, and draws valid conclusions about the antimicrobial products evaluated.[78] When one designs a clinical trial, it is critical that the research objectives be explicitly stated, and that the study be designed to achieve those objectives clearly, concisely, and unambiguously. This is the key to a successful evaluation program and it has been our experience that this approach appreciably expedites FDA review.[32] Now, let us look at how a study is designed based on clearly defined study objectives.

Simply stated, the study objectives are to provide answers to questions the study is intended to address. These may be the easiest to draft of all the necessary documentation relevant to clinical trials, but they are also among the most important. Every other document or concern of the study will be in support of the objectives. Given the objectives, the general schema will be to design the study relative to those objectives and, with statistical validation integral to the process, to conduct the study employing appropriate microbiological methods and, finally, to evaluate the results applying predetermined, appropriate biostatistical methods. To accomplish this efficiently, several conceptual components should be clearly conceived. These include descriptions of the study's purpose and the experimental strategy constructed to achieve the objectives, including microbiological methods applied, and statistical methods used to evaluate the data.

A.   Description of the Study's Purpose

A concise description of the study's guiding purpose is critical.[79] This point is so implicitly obvious that it is often explicitly ignored until the entire study has been completed, by which time the original objectives have become obscure and unclear. Then, the investigator must backtrack through the study to determine what the original objectives were and, all too often, attempt to make the data relevant to these.

In determining the study's purpose, the three parameters of antimicrobial efficacy measurement (the immediate, persistent, and residual effects) must be specifically considered.[32] If one is evaluating a pre-venipuncture antimicrobial product, the purpose will probably only address the immediate antimicrobial effects. However, if one is evaluating a surgical scrub product, the immediate, persistent, and residual antimicrobial effects will be of importance. It is also important that an appropriate reference (control) product be used in the study as a gauge for comparison of the total antimicrobial effects and to provide an internal validation of the study.

B. Description of the Experimental Strategy

Once the study's purpose has been determined, one can begin designing the study to meet that purpose.[32] The design of the study must be such that the study's objectives and purpose are achieved and based on valid results. Validity in experimental design is concerned with two areas: internal and external validity.[33]

1. Internal Validity

Usually, the internal validity of a study can be assured by using the appropriate statistical experimental design.[14] For example, internal validity can be built into the study by random sampling, by blocking, and by the use of specific statistical and experimental controls, as appropriate.[80] There are various errors that can cause a study to be internally invalid. Two of them are routinely encountered in clinical trials.[81] These are as follows.

1. *History*: the specific events occurring between the first and subsequent measurements. An example in clinical trials could be subjects washing their skin surfaces with antimicrobial compounds between the baseline and posttreatment measurements, causing the posttreatment measurement to be biased by the uncontrollable effects of extraneous antimicrobial compounds. The antimicrobial product actually being evaluated potentially would rate higher in efficacy than it actually is.

2. *Instrumentation*: changes in the observations or measuring instruments (e.g., agar plates) may produce changes in the study results.[32,81] An example in clinical trials could be the use of different media lots that are significantly different in nutritional characteristics and, therefore, are variable in their ability to support microbial growth. Another example is studies for which subjective evaluations must be provided, like those measuring healthcare personnel skin irritation after use of

various handcleansing products. Here, two observers may score differently a degree of chafing or redness that is, in fact, the same.

2. External Validity

Unfortunately, no experimental designs have built-in controls for threats to external validity.[32,81] External validity refers to the extent to which the results of a specific study can be generalized to the population-at-large or to other environmental conditions. External validity can be subdivided into these two perspectives: population validity and ecological validity.[81,82] Population validity deals with the generalizability of results of a study to the general population, whereas ecological validity concerns the generalizability of results of the study to other settings.

The most obvious way of assuring the external validity of the study is to have the identical study conducted independently at different geographic locations and/or among different subject populations.[82,83] If the same results are obtained and the same conclusions drawn by different investigators at each site and/or among each population, the external validity of the study can be considered, to a high degree of probability, satisfactory.

With these thoughts in mind, let us return to the surgical scrub evaluation and some of the mechanics necessary to performing a successful evaluation. Recall that we are interested in the immediate, persistent, and residual antimicrobial effects of the test and control products. A completely randomized design should be devised relative to assignment of the products. That is, each subject in the study must be equally likely to be assigned the test or control products to be used in the study.

In order to compare the two products for their immediate, persistent, and residual antimicrobial effectiveness, we need some value against which to assess any postwash reduction in microorganism populations. That value is a baseline value, the microbial population numbers that normally reside on the surface of the hands, determined by sampling prior to product-testing.

In this study, the immediate antimicrobial effects will be measured just after the surgical scrub, and the persistent antimicrobial effects will be measured at both 3 and 6 h after the scrub. The residual antimicrobial effects will be measured over the course of 5 consecutive days of product use. Once again, by taking the time to sketch out basic experimental design procedures, one can clearly and simply address the requirements necessary to achieving the study's objectives.[32] This is represented using the following experimental design, which is a pretest/posttest design (Table 2). Schemata of this type can be designed to accommodate any number of test products and are crucial to accurate conceptualization of what will actually happen in the testing process.[32]

It is also important to know the distribution of the resulting data.[18] Recall

Table 1 Experimental Design for Surgical Scrub Evaluation

	Baseline measurement	Independent variable (treatments)	Dependent variables (microbial counts over both hours and days)		
R_A	O_1	A	O_2	O_3	O_4
R_B	O_5	B	O_6	O_7	O_8

where:

R = Completely randomized design; each subject is equally likely to be selected into group A or B

O_1, O_5 = Dependent variables (microbial counts during baseline)

Independent variables:

A = Treatment with test product (group A)

B = Treatment with control product (group B)

O_i = Dependent variables (microbial counts after treatment with product A or B).
$O_i = O_2, O_3, O_4, O_6, O_7,$ or O_8

that the dependent, or response variable is the variable one must measure after all other variables have been controlled. It is more often known, simply, as "the variable." Typically, in a topical antimicrobial evaluation, the response variable consists of the microbial colony counts. A problem presented in the statistical analysis of microbial colony counts is that the data collected are routinely nonlinear, and are instead, exponential. Since the vast majority of statistical models are linear models, they cannot be used to evaluate nonlinear data.[85] Therefore, the microbial counts must be transformed to a linear scale (most often, a $\log_{10}$ scale) to be evaluated statistically.

It is also necessary in statistical designs to establish the levels of both the alpha (α) and the beta (β) error, so that the appropriate number of subjects to be tested and the number of replicate measurements to be taken in each sampling period, relative to the desired confidence level, can be determined[32,75]. Recall that α-error (Type I error) is committed by rejecting a null hypothesis that is true, and β-error (Type II error) is committed by accepting a null hypothesis that is false. In other words, α-error occurs when one states that there is a difference between products, when there actually is not, and β-error occurs when one concludes that there is no difference between products, when there actually is. The easiest way to control both α- and β-errors is to use enough subjects (replicates) such that the possibility of both α- and β-errors is reduced. Otherwise, merely adjusting the α-error to a very small level for significance will only serve to increase the probability of β-error.[2,80]

Often the question arises in topical antimicrobial evaluations whether the study should be single- or double-blinded. Recall that in single-blind situations, the participants in the study do not know if they are receiving the test product(s) or the control product, but the investigator does. In a double-blind study, neither the investigator nor the subjects know who is receiving which product. In the "real world," pressure on an investigator to evaluate the sponsor's product in the best light can impose a tacit biasing of the study. The use of a double-blind study can eliminate this biasing problem. Once the study is complete and the data are evaluated, the true product identities may be safely revealed. For even greater protection from experimental bias, the various processes of the study can be compartmentalized so that no one person has total control of all aspects of the study.[32,86] This is especially important in studies where there is double-blinding, but the products to be evaluated are easy to distinguish. This would happen, for example, when the test product is a chlorhexidine gluconate product and the reference product is an iodophor. It is very easy to distinguish the two, no matter how they are labeled.

Commitment to creating an efficient, statistically-based experimental design will serve to provide a cost-effective model that will generate valid results. For this reason, it will well worth the time and effort directed at perfecting design *before* commencing work on the study.[33]

II. THE RESEARCH SITUATION

A. Effects Influencing both the Subjects being Evaluated and the Technicians doing the Evaluation

1. The Hawthorne Effect

In experiments involving human subjects, a great many subtle influences can distort research results.[87] If individuals are aware that they are participating in an experiment, for example, this knowledge may alter their performance and, thereby, invalidate the experiment. Studies carried out at the Hawthorne, Illinois plant of the Western Electric Company first called attention to some of these factors. In one of the studies, the illumination of three departments in which employees inspected small parts, assembled electrical relays, and wound coils was gradually increased. The production efficiency in all departments generally increased as the light intensity increased. Experimenters found that upon decreasing the light intensity in a later experiment, however, the efficiency of the group continued to increase slowly, but steadily. Further experimentation with periodic rest periods and variations in the length of working days and weeks were also accompanied by gradual increases in efficiency, whether the changes in working conditions were for the better or for the worse. Researchers concluded that the attention given the employees during the experimentation was the major factor leading to the production gains. This phenomenon has since been referred to by psychologists as the *Hawthorne Effect*. The factory workers who carried out the same dull, repetitive tasks month after month were stimulated and motivated by the attention and concern for their well-being displayed by the research workers. A new element had been added to their uneventful existence—not the illumination or the other variables that the researchers were studying, but the researchers, themselves.

The term *Hawthorne Effect* has come to refer to any situation in which the experimental conditions are such that the mere fact that the subject is aware of participating in an experiment, is aware of the hypothesis, or is receiving special attention tends to improve performance. Certainly, many educational experiments report changes and improvements that are due primarily to the Hawthorne Effect. For example, research in which one group of teachers continues with the same teaching methods they have previously employed, while another group is trained in a new method, will usually result in changes in teacher performance or student achievement favorable to the new method. Many school districts, in the process of trying out new methods, set up one-year experiments in which a new method is introduced to a limited number of pupils. The result of such experiments are almost certainly influenced by the Hawthorne Effect, because teachers usually approach a new method with renewed enthusiasm and, the students, aware that they are being taught by a new and different method, are likewise likely to display more interest and motivation than usual. The influence of the Hawthorne Effect can be expected to decrease as the novelty of a new method wears off, and therefore, studies extending over a period of two or three years are generally more reliable for evaluating the effectiveness of a new technique.

The most common research strategies employed to assess the magnitude of the Hawthorne Effect (and thereby neutralize its effect on the experiment) employ various kinds of control groups.[88] For example, suppose that a control group H is added for whom such variables as time, effort, interest, attention, novelty, and awareness of participation are essentially equal to those experienced by the experimental group X. Then the difference in outcome of the dependent variable between control group H and a nontreatment control group C would represent Hawthorne Effect, and the difference between control group H and experimental group X would represent treatment effect with Hawthorne Effect removed.

It is interesting that attempts to manipulate the Hawthorne Effect experimentally often have failed to produce evidence of the effect. Nevertheless, there is much indirect evidence that the effect operates in some studies.[32] The prudent researcher should take steps to reduce the special attention given the subjects, the novelty of experimental treatments, and the awareness of participation in a research project, all circumstances that may contribute to the effect. In any event, such precautions will improve the research design, whether the Hawthorne Effect might influence the study's outcome or not.

2. The John Henry Effect

The legend of John Henry tells of an African-American railroad worker who pitted his strength and skill at driving steel railroad spikes against a steam driver that was being tested experimentally as a possible replacement for the human steel drivers. The *John Henry Effect* refers to a situation often found in education research in which a control group performs above its usual average when placed in competition with an experimental group that is using a new method or procedure that threatens to replace the control procedure. This phenomenon is probably quite common in educational studies in which a conventional teaching methodology is being compared with a new methodology.[6] Teachers in the control group feel threatened by the new methodology and make a strong effort to prove that their way of teaching is as good as the new method.

This effect was named and described by Robert Heinrich in 1970 while reviewing studies that compared television instruction with regular classroom teaching. He found that the classroom teachers in the control group often made a "maximum" effort to ensure that their students' performance matched that of students who viewed televised instruction.[8]

The John Henry Effect in human studies probably reflects in part the competitive desire to prove that "I can do just as well as those people who are being trained." It is also probable that persons who know that they are members of a control group feel psychologically threatened by a situation in which they perceive they are "preordained" to come out second best.

Since Heinrich's work, several studies have been conducted in which the John Henry Effect appears to have operated, because unusual effort in the control group was documented, and control subjects matched or exceeded the performance of experimental subjects. Gary Saretsky, who conducted extensive study of this phenomenon, concluded that the John Henry Effect is likely to occur when an innovation is introduced in such a manner as to be perceived as threatening to jobs, status, salary, or traditional work patterns. Saretsky provided convincing evidence that the John Henry Effect resulted in a marked increase in achievement in control group labs when those labs were compared with labs in which performance-contracting was employed.[89] He obtained data on performance of the control subjects for two years prior to the experimental year. These data showed that, during the experimental year, control group gains in laboratory skills, as measured by standardized tests, were much higher than in the two preceding years. Because performance-contracting is very threatening to managers, it seems obvious that the managers made a very strong effort during the year of the experiment.

Obviously, the John Henry Effect could be easily confused with the Hawthorne Effect.[89] The two have somewhat opposite effects on an experiment, however, because the Hawthorne Effect reflects the impact of being part of an experiment upon the experimental group's performance, whereas the John Henry Effect reflects the impact upon the control group in experiments where the experimental group is perceived as competing with or threatening to surpass the control group.

3. The Pygmalion Effect

This effect takes its name from a controversial study by Robert Rosenthal and Lenore Jacobson, reported in *Pygmalion in the Classroom.*[90] These researchers demonstrated that teachers' perceptions of a student's intelligence in some cases appeared to bring about changes in that student's intelligence test scores. Thus, the term *Pygmalion Effect* has come to refer to changes in a subject's behavior that are brought about by the experimenter's expectations. The effect has been replicated in some studies, but not in others. In any case, the possibility that this effect can occur should alert researchers to the importance of *not conveying their expectations* to the subjects. This would likely have a strong effect on the performance of laboratory technicians.

III. MICROBIOLOGICAL METHODOLOGY

As previously discussed, the investigator must be familiar with the microbial ecology of the skin to conduct these types of clinical studies properly. The skin surfaces provide a unique habitat for microorganisms, and knowledge of histology, physical features, and nutrient factors of the skin is important.

The knowledge of its histological structure can aid in understanding both the physical and nutritional characteristics of the skin relative to distribution of microorganisms.[32] The various surface features of normal skin that must be taken into account relative to microbial populations expected at specific anatomical skin sites include eccrine sweat glands, sebaceous glands, apocrine glands, and hair.[14] Physical characteristics of skin influencing microbial growth include pH, temperature, moisture, and the oxygen/carbon dioxide tension of the surface and within pores. Factors that must be taken into account from a microbial nutritional perspective include the host's age, sex, race, diet, nationality, and body morph (e.g., obesity), inasmuch as these influence both surface features and physical characteristics of the skin. Finally, a knowledge of those microorganism types that normally inhabit and colonize the skin surfaces is valuable to the investigator.[32] Although species commonly encountered will vary according to the host variables mentioned, as will their relative numbers and distribution on the host, the normal flora usually includes the coryneform bacteria; the Micrococcaceae, including *Staphylococcus* species and *Streptococcus* species; various Gram-negative bacteria, including *Escherichia coli*, *Mycoplasma* species, dermatophytic molds and yeasts, and viral particles, often transiently. Within this context, the investigator must then determine the best culture medium on which to grow the microorganisms that will be encountered at the anatomical sites sampled. The dilution levels that will be employed in testing and the appropriate incubation temperatures and periods must be established. Again, these kinds of questions must also be addressed before beginning the study.

In the design of topical antimicrobial evaluations, the intended product use must be determined and in-use conditions simulated, insofar as possible.[14,19] For example, if one wants information concerning the antimicrobial properties of a new surgical scrub product, inoculating the hands with an indicator microorganism such as *Serratia marcescens* may not be the most valid approach. Assessment of reductions in normal flora residing on the subject's own skin surface is probably better, since it more closely approximates the actual conditions.

IV. STATISTICAL METHODS USED TO EVALUATE THE STUDY

It is critical that statistical models (linear regression, analysis of variance, analysis of covariance, Student's *t*-test, or other) be selected to comply with the experimental design.[32] If the investigator does not do this beforehand, it has the same effect as noncompliance to any protocol and may invalidate the entire study.

Although the exact statistical model to be used within the framework of the experimental design often depends on the distribution of the data generated (normal, skewed, bimodal, exponential, binomial, or other), the use of exploratory data analysis (EDA) can help the investigator not only select the appropriate statistical model, but also develop an intuitive "feel" for the data before the actual statistical analysis occurs.[91] The investigator actually sees the shape of the data distribution in terms of graphic displays, can identify any trends that may not be obvious from the unprocessed data, and on that basis, can select an appropriate statistical model.

It is also very important to identify the data-handling procedure in the study design. For example, if a prewritten statistical software package is not being used, the required statistical computer programs must be written, as well as validated. If prewritten software packages, such as Statistical Package for the Social Sciences (SPSSx), Biomedical Data Program (DMDP), or Statistical Analysis System (SAS) are used, the software program must be configured. Each program requires some careful thought in that the configuration must reflect how the samples will be taken, how data will be blocked, and what contrasts will be used.[32] In addition, the data input arrays must be considered to assure that the keyed input is compatible with the software's input capacity. Understandable graphic display formats will aid in interpreting even the most complex data analyses.

V. NEW PRODUCT DEVELOPMENT APPLICATIONS

Now, let us change our focus to the actual development of new products for the antimicrobial market. We will begin with one of the most common problems in the industry, that of attempting to produce one antimicrobial product to meet multiple needs.[14]

A common practice of manufacturers in formulating topical antimicrobial products is to recommend their product as a surgical scrub, as a preoperative skin preparation formulation, and as a healthcare personnel handwash. The reason for this is simple. It takes less research and development work to offer one product covering all three categories than to develop a specific surgical scrub product, a specific preoperative skin preparation formulation, or a specific healthcare personnel handwash product.

There is a fundamental error in this rationale, however. Each product category has its own unique requirements, and the topical antimicrobial product should be designed with these requirements in mind, specifically for its intended purpose.[32,33] For example, surgical scrub products, in general, are too harsh and potentially irritating to the skin to serve well for the large numbers of required, repeated washes necessary for healthcare personnel over the course of a day. A surgical scrub product must remove not only transient microorganisms, but also resident microorganisms.[33] The main function of a healthcare personnel handwash formulation is removal of the transient microorganisms. The surgical scrub formulation, then, is "stronger" than necessary to meet the requirements of a good healthcare personnel handwash formulation. Although surgeons may scrub their hands two or three times a day in the surgical setting, it is not uncommon for healthcare personnel to wash their hands 25–30 times, as they interact repeatedly with patients. Used 25–30 times daily, a formulation designed as a surgical scrub frequently produces significant skin irritation and, for this reason, will likely not be marketable.[14] Hence, this market is vulnerable to a manufacturer who will develop a formulation specifically targeted to meet the demand for an efficacious healthcare personnel handwash.

To be marketable, the healthcare personnel handwash product must be effective in reducing transient microorganisms, and it must be nonirritating after repeated and prolonged use. Healthcare personnel have demonstrated a strong preference for products that are mild and nonirritating to their skin, as well as effective.[32] Generally, mildness can be built into the healthcare personnel hand-wash in three ways:

1. By proportionally reducing the amount of active ingredients such as chlorhexidine gluconate, iodophors, or alcohol contained in the product. For example, instead of using the customary 4% level of chlorhexidine gluconate (CHG) found in many surgical scrub formulations, a 2% concentration or less of chlorhexidine gluconate is used.
2. By adding skin conditioners or emollients to the formulation, thereby counteracting the irritating effects of the active compounds and making the product more gentle and mild to the skin.
3. By using a combination of these two methods; that is, a reduction in the active ingredient levels and the addition of emollients and skin conditioners to the formulation.

Often, suppliers are not the manufacturers of the healthcare personnel products they distribute. As a result, they often feel that they are at a disadvantage in that they must "take" what products are offered to them by the manufacturers. The suppliers have more options than often realized, however, even when using preexisting formulations. For example, they could collect samples from several manufacturers for use in a pilot study to determine the product most suitable for their needs. Through this approach, the optimum antimicrobial formulation—low in irritation potential, as well as antimicrobially effective—could be readily identified.

Suppose you want to evaluate both the 2% and the 4% CHG-based formulations of four different manufacturers. An efficient statistical evaluation can be devised that will provide not only the information critical for determining the antimicrobial efficacy, but also the irritation qualities of the products. A potentially useful basis for evaluation of irritation potential is the repeated scoring of skin condition in terms of edema, dryness, erythema, and skin eruptions, perhaps employing a 4-point rating system such as that shown in Table 3. A statistical model could then be designed based on these data to compare the irritation potential of the products.

A particularly useful nonparametric statistical model used for irritation evaluation is the χ^2 (Chi Square) statistic.[82] This model allows the detection of significant differences in the irritation produced by use of the products. The χ^2 statistic is statistically robust (reliable), yet accurate and precise in detecting true significant differences between products, as expressed by ordinal scale data produced from subjective evaluation of skin condition.

Let us now turn our attention from the specific example of evaluating a healthcare personnel handwash to the more general statistical strategy of evaluating new product formulations for which no known antimicrobial performance characteristics are available to the investigator.

Table 2 Rating System for Evaluation of a Product's Skin Irritation Potential

Erythema	0 =	No reaction
	1 =	Mild or transient redness limited to sensitive areas
	2 =	Moderate redness persisting over much of the product-exposed area
	3 =	Severe redness extending over most or all of the product-exposed area
Edema	0 =	No reaction
	1 =	Mild (just perceptible) or transient
	2 =	Moderate-definitely palpable
	3 =	Severe
Rash	0 =	No reaction
	1 =	Mild-few; small eruptions
	2 =	Moderate; scattered eruptions, more than ten per hand
	3 =	Severe
Dryness	0 =	No reaction
	1 =	Mild-transient, generally limited to cuticles, knuckles
	2 =	Moderate; persistent, extending over much of the hand
	3 =	Severe; persistent, extending over most of the hand, characterized by cracking and roughness

A. Marketing Concerns

Although this is not often considered in the research and development effort, it is critical. The most effective topical antimicrobial product is not of much value if there is no market. So the question becomes, "What are the market needs?"[32]

The topical anti-infective market, which mainly comprises healthcare personnel handwashes, surgical handscrubs, and preoperative skin preparation formulations, is experiencing "new" product developments on an almost monthly basis. A casual overview of the current advertising literature demonstrates this. But after critical review, are these developments really focused on "new" products or are they the standard products delivered in a new manner? The answer is "yes," to both these questions. There are new products being developed for the anti-infective market, as well as new systems of delivery and new users for standard products.

Let us now take a closer look at the research and development activity in the surgical scrub and preoperative skin preparation markets. Since we have previously discussed the healthcare personnel handwash formulations in some detail, it will be omitted from this section.

1. Surgical Scrubs

Recall that surgical scrub formulations are designed to remove both the transient microorganism population, as well as a large proportion of the normal endogenous microorganism population residing on the hands. To be considered effective, surgical scrub formulations must demonstrate both immediate and persistent antimicrobial effectiveness (up to six hours postscrub) and, optimally, demonstrate a "residual effect" by becoming more effective antimicrobials with repeated use over time because the active ingredient(s) are absorbed directly into the skin.[2]

Over the years, the povidone-iodine marketshare has been significantly eroded by chlorhexidine gluconate formulations because of the latter's residual properties and, recently, there has been interest in developing low-level (2 or 3%) chlorhexidine gluconate formulations for the surgical scrub segment. Although, at the time of this writing, there were no low-level chlorhexidine gluconate surgical scrub formulations approved for use by the Food and Drug Administration, they have shown impressive antimicrobial performance in clinical trials.

The bulk of the surgical scrubs marketed in the early 1990s were 4% chlorhexidine gluconate products. The market is also likely to see the introduction of several low-level chlorhexidine gluconate products that are effective antimicrobial solutions, but less irritating to the skin.

2. Preoperative Preparative Solutions

Recall that the preoperative skin preparative solution is designed to both degerm an intended anatomical surgical site and provide a high level of persistent antimicrobial activity (up to four hours postpreparation).[32] In the past, preoperative skin preps have been the domain of the iodine products, but these are being aggressively challenged by several chlorhexidine gluconate formulations. In addition, formulations providing long-term persistent antimicrobial activity (up to 96 hours after preparation) are likely to be introduced.

There is also interest in developing new product delivery systems. For example, one company recently launched a highly successful iodine-alcoholic film barrier system. The active iodine ingredient is only 0.5% and, yet, it is very effective in preventing microbial recontamination of the prepared site by its patented "film" barrier system. Other new delivery systems that effectively dispense preoperative preparative formulations to the intended surgical site, but with a "no-mess application," will most likely be introduced.

3. Other Considerations

Although there is some market activity in developing and introducing new active ingredients for the healthcare personnel handwash, surgical scrub, and preoperative skin preparative products, most of the efforts will be focused on using the common antimicrobials (e.g., iodophors and chlorhexidine gluconates) as the active ingredients, but applied in new and novel delivery systems.[2]

Healthcare personnel formulations will increasingly be developed as healthcare personnel products, not relabeled surgical scrubs. Surgical scrub formulations introduced in the 1990s will, again, tend to be modifications of existing products. The preoperative skin preparative solution area will experience the most activity, with new product delivery systems.[2]

Also of note, several new applications will be developed and introduced for existing products. One will be a "full-body shower wash" to be used in conjunction with preoperative preparation regimens.[15] The full-body shower wash will be employed to reduce the microbial baseline counts on the patient's body before being prepared for surgery. The preoperative skin preparation formulations will then have a reduced microorganism population to contend with, making it more effective in its intended use.

We are often asked, "How does one market topical anti-infective products?" But there is no esoteric "trick" to successful marketing of topical anti-infective products; only thorough planning and implementation of a creative marketing program are required.[92] An analysis of new product development programs would generally show that the programs themselves are well thought-out and ultimately very successful. Frequently, however, these programs are based on good intentions, but do not meet the actual market requirements.

Not uncommonly, companies will have problems when they enter the topical antimicrobial market with a product that is essentially the same as a competitor's product and, in doing so, face an uphill battle trying to market a product that is not unique.[22,92] There are many suppliers of surgical scrub, healthcare personnel, and preoperative skin preparative formulations, and most of their products are basically identical to one another. For example, the vast number of CHG products used for surgical scrub formulations are essentially identical 4% CHG solutions. Since there is no real difference between these products, the major selling point ultimately becomes the sales price.

This situation is particularly unfortunate when the industry has such tremendous potential for new, innovative products, as well as new applications for existing ones. For example, there are several alcohol-based products that do not need a water rinse. There is also a need for procedures to be used in conjunction with the preoperative skin preparative process to enhance the total antimicrobial effect. One promising procedure of this kind is the full-body shower wash designed for use by patients at home before elective surgical procedures.[15] Another valuable adjunct to current process would be a long-acting, broad-spectrum topical antimicrobial product for use with patients requiring long-term, venous catheterization. These and many more creative new applications are awaiting development by innovators in the industry.

It is important to appreciate that when new product evaluation methods are designed for new products and receive the FDA stamp of approval, these methods will very likely be adapted industry-wide as the procedures for evaluation of future "me, too" products. If a full-body shower wash product application significantly reduces the microbial flora residing on the skin, the follow-up preoperative skin preparation procedure will likely be more efficacious than it would be alone, since fewer microorganisms with which to contend are left. Hence, a new standard procedure for presurgical prepping would evolve that employs the full-body shower wash and the preoperative skin preparative procedures in elective surgery.

VI. CONCLUSION

We have concentrated on a broad, integrative perspective to evaluating topical antimicrobial products in this chapter. We have discussed these evaluations from an experimental design perspective, a microbiological perspective, a biostatistical perspective, and a marketplace perspective. It is my belief, based on experience, that committing to such a holistic, detailed approach to program design significantly aids in evaluating products, and in moving successfully through the FDA approval process because of the straight-forward, concise, and unambiguous manner of conducting valid clinical trials of topical antimicrobials.

6

Epistemological Requirements in Evaluating the Effects of Specific Treatments

It is important to clearly have in mind the purpose of an evaluation. What is it one is attempting to evaluate? For example, if a specific therapy is claimed to alleviate an "allergy," it is important to define precisely what is meant by "allergy." Does the therapy treat allergic symptomology, the immune response effects of B-cell immunocytes secreting immunoglobin "E," or both? Which quadrants are affected? Additionally, what does alleviate mean? Does it mean "in terms of clinical reduction of IgE in the blood stream" and, therefore, focus on histamine reduction? Does it mean subjective relief from allergies by a patient? Perhaps both. Whatever it is, it is critical that one articulate this precisely.

I. EXPERIMENTAL STRATEGY

Once the purpose of the inquiry has been determined, one must decide how to evaluate a specific treatment modality. The data collected may be qualitative, such as a person's perceptions, feelings, beliefs, experiences, or goals. Or they may be quantitative measurements such as blood chemistry, blood gas concentrations, EEG brain wave patterns, and EKG heart function patterns. It does not matter. The important point is that the conclusions reached are drawn based on the experimental results, that is, the conclusions are grounded in the data. The main experimental study strategy is to assure that the conclusions reached are valid from both an internal and external perspective.

II. CHOOSING QUALITATIVE OR QUANTITATIVE METHODS

When one considers evaluation design alternatives, the strengths and weaknesses of qualitative and quantitative data must be taken into account. Qualitative methods permit the investigator to study the clinical treatment in depth and in detail, without being constrained by predetermined scales for measurement of treatment efficacies.[75] Quantitative methods require the use of standardized measures since they focus on objective, readily observed parameters (blood pressure, EEG for brain function, EKG for heart function, etc.).[76]

An advantage of a quantitative approach is that it allows measurement of treatment responses in such a way as to provide linear data or data that can be linearized, a basic requirement of the statistical models most useful in testing quantitative data.[18] This enables the investigator to obtain a broad, generalizeable set of findings that are both succinct and parsimonious. Conversely, qualitative methods typically produce much more detailed information about smaller groups of patients and treatments.[75] This can provide understanding of the cases and situations studied, but at the cost of reduced generalizability.

Validity in quantitative investigations often depends crucially on careful instrument, construction, and operation that assure the instrument measures what it is supposed to measure. The focus, then, is on the measuring instrument that evaluates the effect of the treatment modality. In qualitative evaluative designs, the investigator is the instrument. Validity in qualitative studies thus depends, to a great extent, on the skill, competence, and rigor of the person doing the study.

Since qualitative and quantitative methods have different strengths and weaknesses, they offer alternative, but not mutually exclusive, strategies for evaluating various treatment modalities. For many studies, both qualitative and quantitative methods, or either, can be used to evaluate the same data.

III. STATISTICAL METHODS

The vast majority of quantitative research designs utilize biostatistical methods to evaluate the outcomes in studies that assess therapeutic modalities from a general or group perspective. It is critical to select appropriate statistical models (e.g., linear regression, analysis of variance [ANOVA], analysis of covariance [ANCOVA], Student's t-test, or others) that utilize the resultant data effectively, keeping type I and II errors to a minimum. Recall that α-error (type I error) is committed when one rejects a true null hypothesis, and β-error (type II error) is committed by accepting a false null hypothesis.[77,78] In other words, α-error occurs when one states that there is a difference between treatments when there really is not. β-error occurs when one concludes that there is no difference between treatments when there really is. The easiest way to control both α- and β-errors is to use more subjects (replicates) so that the possibility of both α- and β-errors is reduced.[79] Otherwise, merely adjusting the α-error to a very small level will increase the probability of β-error.

If one wishes to evaluate the efficacy of a particular treatment approach through the use of statistics, the approach must be planned carefully to assure that testing of the study data will be valid. This enables the investigator to decide on what data are appropriate for analysis through application of what statistical method. Furthermore, when an investigator collects treatment data that are subject to uncontrollable experimental error, statistical methods provide the only approach which can provide valid conclusions from the data.[32]

There are two components primary to any statistical evaluation: the design of the experiment and the statistical analysis of the data. These are interdependent since the method of analysis depends directly on the design employed.

There are three basic principles in statistical experimental design:

1. Replication
2. Randomization
3. Blocking

Replication means that the basic experimental measurement is made repeatedly in the same units of measure. For example, if one is measuring the CO_2 concentration of blood, the measurement would be repeated under similar experimental circumstances. Replication serves several important functions. First, it allows the investigator to estimate the experimental error (standard deviation). This estimate of error becomes a basic unit of measurement for determining whether observed differences in the data are actually "statistically significant." Second, if the sample mean (i.e., $\bar{x}$) is used to estimate the effect of a treatment factor in an evaluation, the replication allows an investigator to obtain a more precise estimate of this effect. If σ^2 is the variance of the data, and there are n replicates, then the variance of the sample mean is $\sigma_{\bar{x}}^2 = \dfrac{\sigma^2}{n}$.

The practical aspect of this is that if no replicates are made ($n = 1$), the investigator may be unable to make a satisfactory inference about the effect of a particular treatment modality. The observed difference could be experimental error and not a true difference. However, if the sample measurement was replicated ($n > 1$), and if the experimental error was relatively small, then the difference observed between treatments ($\bar{x}_1 \neq \bar{x}_2$) could be concluded actually to exist.

Randomization of testing procedures performed on subjects is the mainstay of statistical analysis.[80] No matter how objective an investigator is, experimental bias may creep into the study unless random sampling methods are used. Often, investigators create experimental bias when they try to randomize an experiment subjectively. The best way to avoid this form of bias is through a formal scheme of random sample selection, using a table of random digits.[77] Such tables are available from various sources, including most basic statistical texts and random number-generating computer programs.[77,79,80] Through randomization, both the treatment modality a subject is assigned and the order in which they are evaluated are randomly determined. That is, each subject is equally likely to be selected for a particular treatment at a particular time. Virtually all statistical methods require that the treatment observations (as well as their errors) are independently distributed random variables.[80] By randomization, one can average out the effects of extraneous factors (confounding variables) which are present. For example, in a study using both males and females in a treatment program, randomly assigning subjects to each treatment will average out weight differences of males versus females, which may be important in the dosage level assigned.

Blocking is a technique used to increase the precision of an experimental design.[76] A block is a portion of a treatment evaluation and it involves making a comparison of randomly assigned treatments within each block. For example, in a cross-over design, one-half of the subjects are assigned Treatment A and the remainder, Treatment B. After a "wash-out period," those who got B now get A, and vice versa. Time is the "block" variable in this case.

The exact statistical model selected depends, in part, on the data distribution (normal, skewed, bimodal, exponential, binomial, or other).[18,76,79,80] As will be explained, the use of exploratory data analysis (EDA) procedures can help the investigator select the appropriate statistical model and develop an intuitive "feel" for the data before the actual statistical analysis occurs.[81] Extreme caution must be exercised to avoid using quantitative designs to measure what are qualitative phenomena more appropriately evaluated using various qualitative approaches.

IV. STATISTICAL MODEL-BUILDING AND ANALYSIS

A. Exploratory Phase

Scaled-down pilot studies are very useful in exploring the effects of a certain treatment regimen before expending time/money on a full-scale study.[32] Study objectives, experimental design, and measurement methods can often be adapted from other studies examining a similar independent (or controlled) variable. The dependent or response variable—the treatment effects—cannot be known; hence, exploration of those collected data will be very useful in selecting the appropriate statistical models to evaluate the data as inexpensively as possible. This is done through exploratory data analysis (EDA). Exploratory Data Analysis is used to evaluate the data and to develop an "intuitive feel" for them.[32,81] The process consists of four major facets:

1. *Data displays*, that allow the investigator to observe and intuitively comprehend the data distributions and patterns under study.
2. *Residual displays*, that enable the investigator to fit an appropriate statistical model to the data by interactive statistical model-building.

From the residual values (the difference between the actual data values and the expected numerical data values predicted by the statistical model) as an indicator of the statistical model's predictive ability, the investigator can generate a very effective, reliable statistical model.

3. *Reexpression* often simplifies the statistical analysis by rescaling the data through any of a variety of mathematical restatements, such as reciprocals, square roots, and log_{10} restatements. When working with nonlinear data, such as those encountered in clinical trials, the reexpression procedure serves to linearize the data so a simple linear statistical model can be applied to them, instead of more complex nonlinear functions.

4. *Use of resistant and robust mathematical models*, that are affected little by the extreme data value outliers known to significantly affect parametric statistical models. Use of the robust models often produces more reliable evaluations, even when using very few subjects, as in pilot studies.

Also, by employing EDA, one can often select a statistical model representative of a real-world situation.[14] This makes the statistical analysis of the product's performance more accurate and precise and, therefore, more realistic.

To select the most optimal statistical model—the one that realistically portrays the data—several EDA data-handling procedures and displays stand out They include the following:

1. *Stem-and-leaf* displays, that provide a flexible, effective procedure for ordering raw data and, thereby, determining their distribution (e.g., the normal or a log-linear distribution).

2. *Letter value displays*, which summarize and describe raw data in terms of their dispersion relative to the median value. That is, they show where the values lie, or do not lie, relative to the median and provide insight into the distributional shapes (e.g., skewed left or skewed right) of the data.

3. *Box plots* graphically display values in a format that resembles the standard Student *t*-confidence level diagram. Box plot displays are strictly visual, with no numerical values provided. One can use them to compare groups of data, much like one uses a series of 95% confidence intervals (CI).

Once a good understanding of the data distribution is obtained, a statistical model to be used with an appropriate sample scheme is chosen.

B. Statistical Model Selection

The statistical model, to be appropriate, must measure the data accurately and precisely.[18] The test hypothesis should be stated as clearly and concisely as possible. If, for example, the study is designed to test whether or not products A and B are equivalent over the course of multiple washings, the statistical model should accurately test that hypothesis and measure it.

Roger H. Green,[82] in his books *Sampling Designs and Statistical Methods for Environmental Biologists*, describes ten steps for effective statistical analysis. These steps are applicable to any topical antimicrobial product analysis:

1. State the test hypothesis concisely to be sure that what you are testing is what you want to test.

2. Always replicate the samples. Without replication, measurements of variability may not be reliable.

3. Insofar as possible, keep the number of sample replicates equal throughout the study. This practice makes it much easier to analyze the data and produces more reliable results.

4. When determining whether a particular condition has a significant effect, be sure to take samples both where the test condition is present and where it is absent. (Example: If you find, through analysis, that reduction in a product's active ingredients neutralizes its measured effects, you should demonstrate that this problem can be corrected by increasing the level of active ingredients.)

5. Perform a small-scale study to provide a basis for sampling design and statistical model selection before going to the trouble and expense of analyzing the entire manufacturing system.

6. Verify that the sampling scheme you devise actually results in a representative measure of the population you want to evaluate. Guard against systematic and experimental bias by using techniques of random sampling.

7. Break a large-scale sampling process into smaller components.

8. Verify that the collected data meet the statistical distribution assumptions. In the days before computers were commonly used and programs were readily available, blind assumptions had to be made about distributions. Now it is relatively easy to test these assumptions, at least in part, before accepting the statistical model as valid.

9. Test your model thoroughly to make sure that it is useful for the process under study. And, even if the model is satisfactory for one set of data, be certain that it is adequate for other sets of data derived from the same process.

10. Once these nine steps have been carried out, one can accept the results of analysis with confidence. Much time, money, and effort can be saved by following these ten steps to system analysis.

Once an investigator has a general understanding of the product's attributes, a statistical model must be made to fit the data.[18,79,80] At times, the data are such that nonparametric models are more appropriate than are parametric models. For example, if budgetary or time constraints force the investigator to use only a few subjects per product in a study, or if some requirements of the parametric model, such as a normal (Gaussian) distribution of the data, cannot be achieved, then the nonparametric model is the model of choice.

Prior to mounting the large-scale study, the investigator should reexamine: (a) the test hypothesis; (b) the choice of variables; (c) the number of replicates required to protect against type I and type II errors; (d) the order of experimentation process; (e) the randomization process; (f) the appropriateness of the statistical model used to describe the data; and (g) the data collection and data-processing procedures, to ensure that they continue to be relevant to the study.[32] Also, once again, the product must be marketed to be successful. Therefore, another review of the product in relation to the needs of and competition in the marketplace is in order.[84]

Let us now address briefly the types of parametric and nonparametric statistical models available. More detail on these will be provided in Chapter 8.

V.   PARAMETRIC STATISTICS

Parametric statistics require that certain conditions be met and, generally, these include the following:[18,76,79,80]

1. The observations must be independent. That is, the selection of any one case from the population for inclusion in the sample must not bias the chances of any other case for inclusion, the value that is assigned to any case must not bias the chances of any other case for inclusion, and the assigned score must not bias the score which is assigned to any other case.
2. The observations must be drawn from normally distributed populations (the so-called Gaussian distribution).
3. The populations sampled must have the same variance.
4. The variable involved must have been measured at least on an internal scale, so it is possible to perform arithmetic operations (adding, dividing, squaring, finding means, etc.).

All of these above conditions are requirements of parametric models. Ordinarily, these assumptions are not assured through statistical model checking and validation procedures. Rather, they are presumed to hold. The meaningfulness of the results of a parametric test, however, depend upon the validity of these assumptions.

Parametric statistics, which include the Student's *t*-test, linear regression, analysis of variance (ANOVA) and analysis of covariance (ANCOVA), utilize parameters (mean, standard deviation, and variance). The data collected are termed "interval" data (102.915, 1×10^{-5}, 7.23914 . . .). Interval data can be ranked, as well as subdivided, into an infinite number of intervals that are objectively meaningful. In other words, a scale measurement is applied that is continuous and repeatable, regardless of who takes the measurement; subjectivity, insofar as divisions in the measuring device have meaning, plays no role. Usually, interval data relate to some standard physical measurement (e.g., height, weight, blood pressure, the number of deaths, etc.). Subjective perception of degrees of pregnancy, prestige, social stress, etc. cannot be translated into interval data, despite a number of research designs erroneously categorizing them as such. Extreme caution must be used when using quantitative designs to measure what is really qualitative (noninterval) data.[18,76,79,80]

Common parametric statistical models include the following:

1. *Student's t-test*: probably the most common parametric statistical model is the Student's *t*-test. It is often used to compare two groups of data to each other—for example, to compare those of a test group to a specific value or to compare data from two groups (a test and a control group or two test groups). It can be used as a "one-tail" test to determine if one group is significantly "better" or "worse" than another, or a "two-tail" test to determine only if they differ significantly.

2. *Analysis of variance* (ANOVA): analysis of variance is also a common parametric statistic used to compare more than two groups. There are a number of variants of this model, depending upon the number and combination of groups, categories and levels one desires to evaluate. Common forms of ANOVA include one-factor, two-factor, and three-factor designs, as well as cross-over and nested designs.

VI. NONPARAMETRIC STATISTICS

A nonparametric statistic does not require the four general conditions necessary for use of parametric statistical models with the exception of random sampling.[83,85,86] Nonparametric statistics do not utilize parameters (mean, standard deviation, and variance) and can be applied to evaluate noninterval data—nominal or ordinal data—but interval data as well, if appropriate.

Nominal data can be grouped, but not ranked. Data such as right/left, male/female, yes/no, and 0/1 are examples of nominal data and such data consist of numbers used only to classify an object, person, or characteristic. Ordinal data can be both grouped and ranked. Examples include good/bad, poor/average/excellent, or lower class/middle class/upper class, which require subjective evaluation translated to a numeric scale. Nonparametric statistics also may be applied to interval data when the sample size is very small, and the data distribution cannot be assured to be "normal," a requisite for use of parametric statistics. A normal ("bell curve" or Gaussian) distribution is not a requirement of many nonparametric models.[83]

Common nonparametric models include the following:

1. *Mann-Whitney Statistic*: this test is the nonparametric analog to the Student's *t*-test and it is used to compare two groups to one another. Unlike the parametric Student's *t*-test that assumes a normal "bell-shaped" distribution, the Mann-Whitney statistic requires only that the sample data collected are randomly selected.
2. *Kruskal-Wallis Model*: this is the nonparametric analog to a one-factor ANOVA model and it is used to compare multiple groups of one factor. For example, suppose one wants to evaluate the antimicrobial effects of five different hands soaps, the Kruskal-Wallis Model could be employed for this evaluation.

In many therapeutic evaluations, where the number of human subjects required to perform the study is low and, thus, cost-feasible, the best quantitative choice may well be the nonparametric statistical designs. Most investigators simply will not have the funding available to perform elaborate studies. Additionally, since the statistical error of nonparametric studies will tend to be type II (stating the treatment is not effective, when it really is), when effective treatments are observed, they will tend to be highly significant.[85,86]

VII. QUALITATIVE RESEARCH DESIGNS

Qualitative research designs generally focus on three kinds of data collection: (1) in-depth, open-ended interviews; (2) direct observation; and (3) written documents.[75,79] The data from interviews consists of direct quotations from patients about their experiences, opinions, feelings, and knowledge. The data collected from observations consists of detailed descriptions of patients' treatment activities and behaviors, their actions, and the full range of interpersonal interactions and organizational processes that are part of the human/patient experience. Such data may be gleaned from document analyses that yield excerpts or entire passages from clinical or program treatment records, memoranda and correspondence, official publications and reports, personal patient diaries, or open-ended, written responses to questionnaires.

The data for qualitative analysis typically comes from field work.[88] During the field work, an investigator spends time in the setting under study—a treatment program, clinical setting, retreat, or wherever situations of a treatment modality which is important to a study can be observed and people interviewed. The investigator makes first-hand observations of activities and interactions, sometimes personally engaging in the treatment activities as a "patient observer." For example, an investigator might participate in a treatment program under study, as a regular member, client or patient, talking to people about their experiences and perceptions of the treatment modality, either informally or formally. Relevant records and documents are examined. Extensive field notes are organized into readable, narrative descriptions with major themes, categories, and illustrated case examples extracted through content analysis. The findings, understandings, and insights that emerge from the field work and subsequent analysis represent the value of quantitative inquiry.

The validity and reliability of qualitative data depends, to a large degree, on the methodologies, skill, sensitivity, and integrity of the investigator.[75] Systematic and rigorous observation involves much more than just being present and looking around, and skillful interviewing, more than just asking questions. Content analysis requires considerably more than just reading to see what is there. Generating useful and credible data findings through observation, interviewing, and content analysis requires discipline, knowledge, training, practice, creativity, and work.

Many times, data cannot be presented well in a meaningful, numerical form. This is particularly true for individual and population interior phenomenological designs, where values and meaning are the data and must then be interpreted, not observed.[87] Many of these types of designs measure subjective "quality-of-life" issues. But, they are also useful where there is suspicion that a particular therapy may be beneficial for physiological treatment, but it is not known how to measure the real effects (e.g., visualization of wellness in treating chronic diseases).

Let us now explore several of these qualitative designs—the phenomenological, the heuristic, the historical and the evaluative—in greater detail.

VIII.   PHENOMENOLOGICAL DESIGNS

The method involves questioning participants concerning their perceived experience of the phenomenon under study. Data are usually collected via a tape-recorded, unstructured interview where the participants report all experiences related to the phenomenon. The researcher analyzes these self-reports for essential, recurrent themes which surround the phenomenon. In these studies, the meaning imparted to the experiences is very important. Experience includes sensory, linguistic, numeric, and symbolic content. But keep in mind, it is the meaning that is the central interest to the researcher. If one experiences positive effects from a specific therapeutic modality that may or may not provide positive "physical" therapeutic results, that modality cannot be discarded as "useless" therapy. It may not address physical well-being, but does address quality-of-life issues (emotional, mental, or spiritual) and, therefore, has value. The phenomenological research design provides a means of uncovering the value.

Usually, this method requires a beginning statement by the researcher to orient the study. For example, asking how a particular treatment was experienced can be a beginning statement. The interview is designed to expose recurrent themes in an exhaustive exploration of the phenomenological experience. The analysis is time-consuming and follows a preset sequence. First, tape recordings of the interviews are usually transcribed. The nonessential words are separated from those relevant to the phenomenon. "Meaning units" are highlighted in the transcript. Redundancy is reduced as "meaning units" and taken from collected transcripts of the participants and categorized by the researcher into related clusters or themes. Finally, these themes are reduced to cogent, succinct, one-paragraph descriptions of the phenomenon, as described by the participants. A core theme or multiple themes are the primary findings of this study.

A potential weakness of this method lies with researcher bias of the study. Grouping meaning units, omitting material from the study, and composing descriptions of thematic clusters are all potential sources of bias. Hence, it is important to assure external validity of the study by comparing results with those of others.

IX. HEURISTIC DESIGNS

This research design is used to *discover* the meaning or nature of an experience or therapeutic methods and procedures for further exploration. Discovery is the main focus. This research design uses a series of iterations to ever deepen the exploration process. The results of the first iteration are used as the starting point for the next. It is a common method used by conventional medical practitioners to diagnose disease, as well as to treat it. An example of this research approach may be seen in the treatment of an infection. A practitioner first prescribes treatment A. The results noted from the prescription are the starting point for the next treatment regimen. If no improvement is noted, the practitioner may increase the dose or switch to another regimen and then evaluate its results.

Heuristic designs involve active participation of both the researcher and subject. The researcher continually probes the subject concerning their experience. Probing interviews are common, using a structured or semi-structured interview style. A heuristic investigation will often employ a defined set of questions, based upon a decision tree diagram. The process of data collection is interactive and proceeds in a sequential manner.

When used as a research design, the researcher begins by asking questions concerning the phenomenon of interest. The answer generates more questions.

These questions, in turn, become a catalyst for more extensive inquiry. In research, the investigator questions each of the subjects. The interview direction, of course, depends upon the subject's responses. The investigator compiles a schematic list detailing the various responses of each subject, and then compares and contrasts the responses, generating categories, clusters, and themes from the interviews.

The same general biasing phenomena noted in the phenomenological research study affects the internal validity of this kind of study. To assure external validity, it is important that the research be repeated by other investigators at different geographical locations.

X. HISTORICAL RESEARCH DESIGN

The historical design enables the investigator to perform an evaluation using data collected in the past. The purpose of the historical method is to uncover and organize relevant data from a selected period of time, enabling the researcher to evaluate those data with relevancy to the present. It can be used for interior or exterior phenomena. In this design, the researcher compiles and reviews systematically the documentation of people, places, or events concerning the therapeutic modality under investigation.

For example, if certain individuals are able to avoid the various symptoms of AIDS indefinitely, even though they have tested HIV-positive, an investigator may explore how they have done it. This would involve exhaustive documentation of the many variables, individual or environmental, that could influence progression to active disease. Hence, success in using the design requires careful collection and organization of data, as well as *reliable data*. Historical research designs are often criticized for being subject to high degrees of biasing—from both the investigator and the historical documents used as data. Since the data were collected in the past, there is no way to assure their internal validity. The same precautions listed for use of phenomenological design also apply to the internal validity of this design. To assure external validity, it is important that the research be repeated by different investigators at different geographical locations.

XI. EVALUATIVE RESEARCH DESIGN

This design is used to evaluate different therapeutic modalities for their efficacy in treating a disease "in the field." The research design can either be formative (assessing a therapeutic modality being developed) or summative (assessing a fully developed therapeutic modality). This method requires that the investigator state the therapeutic goal, collect data while the actual therapy is in process, and compare the results of the treatment modality relative to its therapeutic goal. Usually, numerical data are not collected, but rather the qualitative findings of the investigator. An example of this type of research design would be determining the therapeutic benefits of "play" in cases of mild, but chronic, depression.

The evaluative research method enables the researcher to obtain reliable data, so that a decision regarding the initiation, continuation, or termination of a therapeutic modality can be made. The research must be guided by clear formulation of the program goals and the objectives of the evaluation. To assure both the internal and external validity, the requirements noted in the phenomenological research design also apply to this design.

XII. CONCLUSION

It is important that products (or therapeutic regimens) be evaluated for merit as well as potential problems. There are numerous ways to perform these evaluations, using either quantitative or qualitative research designs, and it is important that investigators be familiar with a large selection of these. This will prevent the researcher, who has only one tool—a hammer—from viewing everything as a nail.

7

Evaluation Strategies and Sample Working Protocols

In this chapter, we will limit the focus to the following three areas:

1. Consumer Antimicrobial Products
   a. Handsoaps
   b. Bodysoaps
2. Foodhandler Antimicrobial Products
3. Medical/Healthcare Industry Antimicrobial Products

I. CONSUMER ANTIMICROBIAL PRODUCTS

This category consists of antimicrobial bodysoaps, handsoaps, towelette hand-wipes, and other associated products. The purpose of these types of products is to reduce transient, potentially pathogenic microorganisms picked up in the environment (e.g., countertops, coins, door handles, etc.). These types of products are also often intended to remove excess normal microbial flora found particularly in moist body areas.[33] The types of clinical trials currently used in evaluating these products include the Healthcare Personnel handwash, the Modified Cade handwash, and the general use handwash for handsoaps, and axilla cup scrub sampling for bodysoaps.[22]

The Healthcare Personnel handwash evaluation, as presented in the Federal Register, is rarely used in its entirety for evaluating consumer hand-cleansing antimicrobial products. The Modified Cade and the general use procedures are the most commonly used. For antimicrobial bodywash products, the axilla cup scrub and the full-body wash evaluations are used.

Let us begin with the procedures most often used for evaluations of antimicrobial handwash products.

A. Evaluating Consumer Antimicrobial Handwash Products

Manufacturers and suppliers of consumer antimicrobial handsoaps must evaluate the antimicrobial properties of their products not only to collect in-house product information, but also to assure that a product performs as expected (i.e., that the product is antimicrobially effective and not excessively irritating to the hands). There are a variety of approaches available for evaluating these products; however, we will focus on antimicrobial efficacy evaluations employing human volunteer subjects, and on two common in vitro evaluations, the time-kill and minimum inhibitory concentration tests.

B. Evaluative Confusion

Methodologies for human clinical trials used to evaluate consumer antimicrobial soaps are ill-defined. This is due mainly to the Food and Drug Administration's (FDA) decision not to provide guidelines in testing these products because it is not convinced that consumers are provided any substantial benefits by using antimicrobial soap products on a routine basis. If guidelines were to be provided, then by implication, it would be assumed that the FDA perceived merit in such products.

While most involved agree that the healthcare personnel handwash evaluation is not appropriate for evaluating consumer antimicrobial soaps, some regulatory groups suggest a "conflict of interest" is implicit in the alternatives proposed by the manufacturers. That is, the soap manufacturing associations do not have an unbiased perspective on performance of these products, a crucial requisite to designing valid testing methods.

C. Antimicrobial Handsoaps and Disease Prevention

Let us step back a moment and discuss aspects of infectious diseases and the role antimicrobial handsoaps play in their prevention. For infectious diseases to occur, the following sequence of five events must take place.[13,30]

1. A person must come into contact with the microorganisms.
2. The microorganisms must enter the person.
3. The microorganisms must spread from the entry site.
4. The microorganisms must be able to multiply within the person.
5. As a result of microbial enzymes and toxins and, to some extent, response, tissue damage occurs.

An effective handwash disrupts the disease process after the first event by removing contaminating microorganisms from the hand surfaces. While not all of the microorganisms can be removed, their population numbers can be reduced below the level required to cause disease in normal humans.[13]

When the anti-infective properties of antimicrobial handsoaps are discussed, it is important to understand what types of microorganisms must be addressed. The microorganisms which normally colonize hand surfaces pose little threat of infectious disease and, in fact, serve competitively to exclude colonization by transients.[32] There are, of course, situations where normal or resident microorganisms cause disease when they are introduced into areas where they are not "normal." An infected cut often is an example of this. However, even in this case, appropriate washing serves to degerm the infected area, cleaning it of dead cells and exudative material.

In the vast majority of cases, the threat of infectious disease is from "transient," pathogenic microorganisms that contaminate and may temporarily colonize hand surfaces. Hand contamination with transients occurs when one comes in contact with substances such as mucous, blood, soil, urine, feces, or food. Given the opportunity, the contaminant microorganisms may then infect the person or be passed on to others via hand contact.[22]

D. Functional Parameters of Effective Antimicrobial Soaps

Two parameters are of primary interest in evaluating antimicrobial handsoaps: their immediate degerming effectiveness and their persistent antimicrobial effectiveness.[32] Recall that immediate antimicrobial efficacy is the measure of a handwash product's effectiveness as a function of both the mechanical removal of contaminating microorganisms in the process of the handwash procedure and the immediate inactivation of microorganisms resulting from contact with the antimicrobial ingredient(s) in the soap. The persistent antimicrobial effectiveness is the product's ability to prevent, either by microbial inhibition or lethality, transient microbial recolonization of the skin surfaces after handwashing.

Accurate and precise measurement of these two parameters is often difficult. It is important then that efficacy evaluations be well-defined and clearly stated before conducting them to assure that the "question" asked will actually be answered with valid data.

E. Antimicrobial Efficacy Studies

To determine what microorganism species are susceptible to the antimicrobial soap, as well as the rates of microbial inactivation, certain in-vitro tests should be conducted.[32] These include time-kill evaluations, minimum inhibitory concentration evaluations, and microbial sensitivity tests. But, in order to determine how effective the handwash is in the "real world," in-vivo human product-use studies—clinical studies—must be conducted.[89]

As noted earlier, study designs routinely utilized for this include:

1. The Healthcare Personnel Handwash Evaluation
2. The Modified Cade Handwash Procedure
3. The General Use Handwash Evaluation

Let us discuss these tests in greater detail. Then, I shall propose an approach to evaluation that I believe is optimal for testing the efficacy of antimicrobial handwash products.

II. HEALTHCARE PERSONNEL HANDWASH EVALUATIONS

This testing approach is utilized when evaluating antimicrobial products used in the medical field by healthcare personnel. Enough subjects are recruited to assure that the sample size is adequate in terms of reducing the probability of type I and II errors appropriate to the statistical models selected, thereby providing a high degree of statistical power (the "power of the test").*

A test product, a vehicle product (test product without the active antimicrobial) and a reference product are customarily used in this evaluation. It is expected that the test product will demonstrate significant antimicrobial properties through elimination of microorganisms by mechanical removal and by microbial inactivation as a function of exposure to the active antimicrobial ingredient.[1] The vehicle is expected to demonstrate the product's ability to remove microorganisms through the mechanical action of the handwash only. The reference product is used to validate the study in that results of using the reference product are expected to be similar to those from other studies in which it was used.

A "wash out" period of at least seven days is necessary, during which time the subjects are not permitted to use antimicrobial products or expose their hands to acids, bases, or any other substances known to affect the microbial populations.

---

*Type I (α) error is committed when one rejects a true null hypothesis and type II (β) error is committed by accepting a false null hypothesis. The statistical power is the ability of the test statistic to detect a true alternative hypothesis. In other words, type I error occurs when one states that there is a difference between products when there really is none. Type II error occurs when one concludes that no difference exists between products when there really is one. The power of the test is the ability of the test statistic to conclude that a difference exists when one really does. The power of the test is very sensitive to the amount of variance that is implicit to the data, the discreteness of the difference sought between the systems evaluated, and the size of the sample population. The smaller the variance, the smaller the difference sought and/or the larger the sample population, the greater the power of the test.

On the test day, subjects are inoculated with *Serratia marcescens*, a marker microorganism strain that produces red colonies when plated on Soybean-casein Digest agar. This allows them to be distinguished from microorganisms normally found on the skin, which will appear white or yellowish on the agar. The hands are next sampled using the glove juice procedure to establish a baseline measurement of the number of bacteria inoculated.

The hands are again inoculated with *S. marcescens*, and then a handwash is performed, using the assigned test configuration—test product, vehicle, or control product. Following this, the glove juice sampling procedure is again employed.

The inoculation/wash procedure is repeated 10 consecutive times. Glove juice samples are taken after inoculation/wash cycles 1, 3, 7, and 10, and the dilutions plated on Soybean-casein Digest agar. The agar plates are then incubated at $25 \pm 2°C$ for approximately 48 hours.

The microbial population counts on the hands after washes 1, 3, 7, and 10 are then compared to the baseline average value. To be acceptable as a healthcare personnel handwash, the product must demonstrate at least a 2 $\log_{10}$ reduction from baseline after the first wash (immediate efficacy) and at least a 3 $\log_{10}$ reduction after wash 10 (persistence efficacy).

A. Modified Cade Handwash Procedure

The number of test subjects recruited varies, but it is common to enroll 55–65 subjects. A washout period of at least seven days is observed employing the same product-use restrictions presented for the healthcare personnel handwash procedure. This is followed by a baseline measurement period.

On the first day of the baseline period, subjects wash their hands 5 consecutive times with a bland soap. Microbial samples are collected from washes 1 and 5, using a "basin wash" procedure. The basin wash procedure requires subjects to wash their hands with the nonmedicated bland soap in a polyethylene container such as a zip-lock freezer storage bag containing 1 liter of sterile water. After washing, the wash water is mixed well, and sample aliquots are plated on Soybean-casein Digest agar. The agar plates are incubated at $30° \pm 2°C$ for approximately 48 hours.

For days 2 and 3 of the baseline period, the subjects wash their hands *ad libitum* with the bland soap outside the laboratory. On day 4, those 45–50 subjects having the highest baseline bacterial colony counts from day 1 remain in the study for completion of day 5; the others are dismissed.

On day 5, subjects return to the laboratory to wash their hands as on day 1, and wash 1 and 5 samples are again collected as previously described. Subjects then continue using the bland soap for all handwashing outside the laboratory on days 6 and 7. Finally, on day 8, subjects return to the laboratory and again wash their hands 5 consecutive times, with samples collected from washes 1 and 5 using the basin wash procedure.

The subjects are then assigned their test handwash soap products. Subjects wash their hands with the product a minimum of 3 times per day outside the laboratory, with at least 1 hour between washes. They also use the assigned product for any additional handwashes, as well as for bathing and showering. This regimen is followed for 11 consecutive days. On day 12, subjects return to the laboratory and turn in their assigned test products. They then wash 5 consecutive times with a bland soap. The samples are then again collected using the sterile basin wash procedure for washes 1 and 5.

Usually, the first wash samples collected from the baseline period are pooled, as are the fifth washes. That this can be done, however, must be verified by proving the data to be pooled are, among them, not statistically different. The count data from the first test sample wash are then compared to the pooled wash 1 baseline data, and the count data from the fifth test wash are compared to the pooled wash 5 baseline data.

B. General Use Handwash Evaluation

Normally, 10–20 subjects are recruited for each test product evaluated. A control product is sometimes included as well. A washout period of at least seven days is observed, during which time subjects are not permitted to use antimicrobial products or expose their hands to compounds known to have antimicrobial properties.[22]

The remaining study procedures are identical to those of healthcare personnel handwash, except that the total number of washes is at least 5, but as many as 10, with glove juice samples collected at baseline, washes 1 and 5, and at wash 10, if 10 washes are used. The microbial colony counts taken from samples after the test washes are compared to the baseline values.

III. PROPOSED ANTIMICROBIAL HANDSOAP EVALUATION

The optimal evaluation design should be practical, yet provide accurate and reliable results, ideally, for reductions in transient microorganisms, not resident ones.[22] The healthcare personnel handwash is designed to evaluate the antimicrobial efficacy of products used in the medical healthcare field and, therefore, is excessively stringent for evaluating consumer-use antimicrobial soap products.

Results generated by Modified Cade Handwash procedure are sometimes considered equivocal, particularly in that the antimicrobial efficacy is evaluated in terms of normal microbial flora data, of questionable value for evaluating consumer antimicrobial hand soaps. Additionally, it measures the residual antimicrobial effectiveness of the test antimicrobial product, not the immediate effects.* Finally, since a control product is rarely used, the validity of any such study is open to question.

The optimal design for evaluating consumer antimicrobial handsoaps is a combination of the healthcare personnel handwash and the general use handwash evaluations. It is based on the removal of transient microorganisms over the course of 5 consecutive inoculation/wash cycles and is practical, accurate and reliable. The proposed study design follows.

A. Test Material

The products to be evaluated are:

Test product:

Lot Number: ⎯⎯⎯⎯⎯⎯⎯⎯
Expiration date: ⎯⎯⎯⎯⎯⎯⎯⎯

Control product:

(test product without antimicrobial compound)
Lot Number: ⎯⎯⎯⎯⎯⎯⎯⎯
Expiration date: ⎯⎯⎯⎯⎯⎯⎯⎯

B. Test Methods

1. Subjects

A sufficient number of overtly healthy subjects over the age of 18, but under the age of 70, are admitted into the study to ensure that at least 15 subjects complete the study per test configuration group. Subjects should be of mixed age and sex and free from clinically evident dermatoses or injuries to the hands or forearms. All subjects must sign Informed Consent Forms prior to participating in the study.

---

*Residual effectiveness is a measurement of the product's antimicrobial effects when utilized repeatedly over time. The antimicrobial is absorbed into the skin and, as a result, prevents recolonization of the skin by microorganisms. It is a measurement more appropriately used in evaluating surgical scrub products, not consumer antimicrobial soap products.

2.   Concurrent Treatment

No subject can be admitted into the study who is using topical or systemic antimicrobials, or any other medication known to affect the normal microbial flora of the skin.

3.   Pretest Period

A 7-day period prior to the test portion of the study will constitute the pre-test period. During this time, subjects must avoid the use of medicated soaps, lotions, deodorants, and shampoos, except as supplied in a personal hygiene kit, as well as avoid skin contact with solvents, detergents, acids, and bases.* They are also instructed to avoid contact with specific kinds of products on a restricted list that is supplied to them (a list should be attached naming any commercial products they will not be allowed to use). Further, subjects must avoid use of UV tanning beds and bathing in pools and/or hot tubs containing chlorine or other biocides. This regimen will allow for stabilization of the normal microbial populations residing on the hands.

Before the initiation of the study, a Study Description and Informed Consent forms must be provided to the subjects. The study must not begin until IRB approval has been obtained for all facets of testing and all relevant material. Trained laboratory personnel should be available to answer any questions which may arise.

4.   Neutralization

A neutralization assay must be performed to assure that the neutralizers employed in the evaluation effectively neutralize the antimicrobial activity of the product.

5.   Experimental Period

Testing will comprise a period of seven days. Each subject will be employed for about 2 hours on only 1 day of that period. A practice wash using a non-antimicrobial bland soap and the same application procedure as prescribed for the test product should precede the actual test portion of this study. It ensures that the subjects understand the wash procedure.

After the practice wash has been completed, a 5.0 ml aliquot of a suspen-

---

*Subjects should be supplied a personal hygiene kit containing a bland soap, shampoo, deodorant and lotion, as well as a pair of rubber gloves. The rubber gloves should be worn when exposure to antimicrobials is unavoidable. Subjects should be instructed to use exclusively the contents of this kit for personal hygiene needs during their participation in the study.

sion of at least 1.0×10^8 cfu/ml *Serratia marcescens* (ATCC #14756, red-pigmented strain) is transferred into each subject's cupped hands. The inoculum is then distributed evenly over both hands and up to the wrist, via gentle, continuous massage for 45 seconds. After a 2-minute air-dry, the glove juice sampling procedure is performed to provide the baseline data. It is followed by a 30-second nonmedicated soap handwash. The temperature of the water used for this and all subsequent wash cycles should be maintained and controlled at 40° $\pm 2°C$.

The microbial inoculum is again distributed evenly over both hands up to the wrist, via gentle continuous massage for 45 seconds. After a 2-minute air-dry, the subjects wash with their assigned product configuration according to the directions supplied. This is followed by the glove juice sampling procedure.

This inoculation/wash procedure is repeated 5 consecutive times, with a minimum of 5 and a maximum of 15 minutes between applications. A transient microorganism sampling of the hands using the glove juice sampling procedure is performed after inoculation/product wash cycles 1 and 5.

6. Glove Juice Sampling Procedure

Following the prescribed wash and rinse, excess water is shaken from the hands and powder-free sterile gloves are put on the subjects' hands. At the designated sampling times, 75.0 ml of Sterile Stripping Suspending Fluid without product neutralizers are instilled into each glove. The wrist is secured and an attendant massages the hand through the glove in a standardized manner for 1 minute. Aliquots of the glove juice (dilution 10^0) are then removed and serially diluted in Butterfield's Phosphate Buffer solution containing appropriate neutralizers.

Triplicate spread plates are prepared from each of these dilutions, using Soybean-casein Digest Agar with Neutralizers. The plates are incubated at 25° $\pm 2°C$ for approximately 48 hours. Those plates providing colony counts between 25 and 250 should be preferentially utilized for enumeration. If no agar plates provide counts in the 25 to 250 range, those plates having counts closest to that range should be used in determining the number of viable microorganisms.

Following the final product application cycle (Application #5) and glove juice procedure, the subjects should be required to wash with 70% isopropyl alcohol for 1 minute, followed by air-dry, and a thorough rinse under tap water in order to eliminate any remaining *Serratia marcescens* from the hands. If desired, a wash using a surgical scrub product may be added as well.

7. Methods of Analysis

The plate-count data collected from this study should be evaluated using standard statistical computer software. The $\log_{10}$ number of viable microorganisms recovered from each hand is designated the recovered or "R" value. It is the adjusted average $\log_{10}$ colony count measurement for each subject at each sampling time. Each R-value is determined using the following formula:

$$R = \log_{10} [75 \times C_i \times 10^{-D}]$$

where:

75 = the amount of stripping solution (ml) instilled into each glove to perform the glove juice sampling procedure.

C_i = the arithmetic average colony-counts for the 3 plates for each subject at a particular dilution level.

D = the dilution factor.

Note: A $\log_{10}$ transformation must be performed on the average plate counts to convert the data to a linear scale. A linear scale, more appropriately a $\log_{10}$ linear scale, is a requirement of the statistical models to be used.

8. Statistical Analysis

A pre/post experimental design is utilized to evaluate and compare the antimicrobial effectiveness of the test product.

Preproduct application	Postproduct application
R A O_{BL}	O_1 O_5
R C O_{BL}	O_1 O_5

where:

R = subjects randomly assigned to the study

A = Independent variable: Test product

C = Independent variable: Control product

O_i = Dependent variable: Microbial counts at baseline (BL) and after product-use washes 1 and 5

Exploratory Data Analysis should be performed on the data to assure they meet the requirements of the statistical models used. Stem-Leaf Ordering, Letter Value plots, and Box Plots should be generated to assure the data collected approximate a Normal (Gaussian) Distribution.[81] If this is the case, a series of Student's *t*-tests can be conducted using the 0.05 level of significance for Type I (α) error. If the data do not approximate a normal distribution, the nonparametric analog of the Student's *t*-test, the Mann-Whitney U test, can be substituted.[33]

C. Comments

It is suggested that the FDA's OTC monograph be followed for the various technical nuances of the study (21 CFR Parts 333 and 369, part III; 17 June, 1994). A minimum of 15 subjects per group is recommended, as is the use of a control product. If one test product is used, then a minimum of 30 subjects should be employed (15 for the test product and 15 for the control product). This is a statistically adequate number, in most cases, to detect true differences between the test and control products, and between the microbial reduction counts from baseline after washes 1 and 5. But this is true only if the standard deviation value for baseline counts is less than 0.5 logs. If the standard deviation is greater, it will be difficult to detect a *true difference* between the test and control products over the two wash samples, and more difficult yet to detect *true microbial reductions* from the baseline values. Hence, using a larger sample size is in the best interest of the product manufacturers, but must, of course, be balanced against cost.

This design for general-use handwash products permits measurement of the immediate degerming effects (first wash) as well as the effects after 5 consecutive inoculation/wash cycles to assure that there is no cumulative build-up of microorganisms after wash sample 5. Additional inoculations/wash cycles and samples will likely add no predictive value to the evaluation.

For this evaluation, the alpha (α) level should be set at 0.05. For screening tests for general product effectiveness, the total number of subjects may be reduced to as few as 10 per group, and the alpha (α) level set at 0.10.

A two-tail Student's *t*-test should be used to assure that data from subjects' left and right hands are equivalent at the 0.05 level of statistical significance for baseline values. If no significant difference exists between the hands at baseline, the hand data can then be pooled.

For simplicity, a series of Student's *t* tests can be used to compare the test and control products to a common baseline value after sample wash times 1 and 5, so long as a correction factor for multiple *t* tests is incorporated. The modified α value ($\alpha*$) must be substituted for the normal α of 0.05.[77] The modified value is calculated as:

$$\alpha* = 1 - (1 - \alpha)^K$$

where:

k   = the number of comparisons made.

α = (normally) the significance level of 0.05.

$\alpha*$ = Adjusted true significance level of all the t-tests used in the evaluation.

Prior to beginning any test procedures, a product neutralizer evaluation must be performed to assure that the active antimicrobial ingredients in the product can be neutralized and to assure that diluents used in the assay have no significant effect on *S. marcescens*. An Analysis of Variance (ANOVA) design should be used to analyze the data from at least 4 configurations:

Configuration 1 (negative control): *S. marcescens* plated on Soybean-casein Digest agar

Configuration 2 (positive control): *S. marcescens* and test product plated on Soybean-casein Digest agar (determines if product kills)

Configuration 3: *S. marcescens* and test product and neutralizer plated on Soybean-casein Digest agar (determines if product has been neutralized)

Configuration 4: *S. marcescens* and neutralizer plated on Soybean-casein Digest agar (determines if the neutralizer affects the growth of *S. marcescens*)

Plate count data from configurations 1, 3, and 4 should be equivalent at the 0.05 level of significance for the neutralizer to be considered adequate.

The log-reduction values that should be required of a general use antimicrobial soap is hotly debated. It is this author's opinion that the reduction values should be at least 1 $\log_{10}$ from the baseline value after wash 1 and 1.5 logs after wash 5.

D. Conclusion

It is necessary that manufacturers of antimicrobial soap products accurately and precisely evaluate and measure the antimicrobial efficacy of their products under simulated in situ wash conditions employing human subjects.

IV. SAMPLE PROTOCOLS (to be modified as required)

Healthcare Personnel Handwash Evaluations (See Medical/Healthcare Personnel Industry Antimicrobials)

The following are sample protocols for commonly used evaluations of consumer antimicrobial soap products. These may be modified to suit the needs of the user.

A. Modified Cade Handwash Evaluation

1. Purpose

The purpose of this study is to determine the effectiveness of one (1) test antimicrobial soap product in reducing the level of aerobic microbial flora on the hands following an extended period of use, using a modified CADE handwashing procedure.

2. Scope

The antimicrobial effectiveness of one test antimicrobial soap product will be determined. The study will begin with 65 human subjects undergoing baseline determination sampling. Those 50 subjects having the highest baseline microbial populations will be utilized in the test portion of the study. After 11 consecutive days of using the test product, the hands will be sampled on day 12, and the microbial reductions from baseline calculated.

3. Test Material

The product to be evaluated and the placebo (nonantimicrobial) soap are:

Test Product
Lot Number: _____
Manufacture Date: _____

Placebo Product
Lot Number: _____
Manufacture Date: _____

4. Test Methods

Subjects. A sufficient number of overtly healthy subjects at least 18 years, but under 70 years, of age will be admitted into the study to ensure that 65 subjects undergo baseline sampling. Those 50 subjects with the highest baseline microbial counts will continue into the test portion of the study. Subjects will be of mixed sex and age. On the day of test, subjects will be free from clinically evident dermatoses or serious injuries to the hands and forearms. All subjects will have the study protocol explained to them and will sign the Informed Consent Forms prior to participating in the study.

5. Concurrent Treatment

No subject will be admitted into the study who is currently using topical or systemic antimicrobials, or any other medication known to affect the normal microbial flora of the skin.

6. Pretest Period

The seven days prior to the test portion of the study will constitute the pre-test period. During this time, subjects will avoid the use of medicated soaps, lotions, deodorants, and shampoos, as well as skin contact with solvents, detergents, acids, and bases. Subjects will be supplied a personal hygiene kit containing nonmedicated shampoo, deodorant, soap, and lotion, as well as latex gloves. Subjects must use only the products supplied in the kit during their participation in the study. They must avoid contact with products on the restricted list. If contact with restricted products is unavoidable, latex gloves will be worn. Subjects will also avoid using UV tanning beds and bathing in pools and/or hot tubs containing chlorine or other biocides. This regimen will allow for stabilization of the normal microbial populations residing on the hands.

Before initiation of the study, a study description will be given to subjects (patient information). Informed Consent Forms, and any other supportive material relevant to the safety of the subjects will be supplied by the principal investigator, following their review and approval by an Institutional Review Board (IRB). The primary purpose of the IRB is the protection of the rights and welfare of the subjects involved in a clinical study (reference CFR 21, Part 56). Trained laboratory personnel will explain the study to each participant and will be available to answer any questions which may arise.

7. Baseline Period

Sixty-five (65) human subjects will be evaluated during the baseline period. The subjects will come into the laboratory on day 1 and wash their hands 5 consecutive times with the Sponsor-supplied placebo soap. The Laboratory Sink Method will be used for washes 2, 3, and 4. For washes 1 and 5, the subjects will employ the Sterile Basin Wash Technique. Samples of the rinsate in the basins will be plated in triplicate onto Soybean-casein Digest Agar plus neutralizers to determine the microbial populations.

On days 2 and 3, subjects will continue to wash outside the laboratory as usual, using only the supplied placebo soap for all normal handwashing and bathing. On day 4, the 50 subjects with the highest microbial counts from their hands will continue into the test portion of the study. The other 15 subjects will be dismissed from the study. All subjects remaining in the study will continue to use the placebo soap for all washing on day 4.

On day 5, subjects will return to the laboratory and again wash their hands 5 consecutive times using the supplied placebo soap. The Laboratory Sink Method will be used for washes 2, 3, and 4. For washes 1 and 5, the subjects will employ the Sterile Basin Wash Technique. Samples of the rinsate in the basins will be plated in triplicate onto Soybean-casein Digest Agar plus neutralizers to determine the microbial populations.

On days 6 and 7, subjects will continue to only use the supplied placebo soap for all normal washing, as needed.

8. Test Period

On the first day of the test period, the subjects will again return to the laboratory and wash their hands 5 consecutive times with the supplied placebo soap. The Laboratory Skin Method will be used for washes 2, 3, and 4. For washes 1 and 5, the subjects will employ the Sterile Basin Wash Technique. Samples of the rinsate in the basins will be plated in triplicate on Soybean-casein Digest agar plus neutralizers to determine the microbial populations.

The subjects will then perform supervised handwashes with their test product at the laboratory starting 30 minutes after the fifth placebo-soap wash. They will wash 3 times with the test product with at least 1 hour between washes. After these washes have been completed, the 50 subjects will be issued test product samples, Laboratory and Home Handwash Log Forms and use instructions.

The subjects will wash 3 times daily at home, each wash no less than 1 hour apart for the test period day 2 and continue washing 3 times daily for the next 9 days (test days 3 through 11). These washes will be documented on the Laboratory and Home Wash Log Forms. Subjects will also use their test product for any additional handwashes, and for bathing and showering.

On evaluation day 12, the subjects will return to the laboratory, turn in their issued products, the Laboratory and Home Handwash Log Forms, and perform 5 consecutive handwashes using the supplied placebo soap. The Laboratory Sink Method will be used for washes 2, 3, and 4. For washes 1 and 5, the subjects will employ the Sterile Basin Wash Technique. The rinsate in the basins will be plated in triplicate on Soybean-casein Digest agar plus neutralizers to determine the microbial populations.

9. Sterile Basin Wash Technique

Before beginning the wash, the subjects will remove all jewelry from the hands and forearms and wet their hands in a 3.79 liter "zip-lock" storage bag containing approximately 1.0 liter of sterile deionized water. The water will be at room temperature (20°–25°C).

Subjects will dispense 2 pumps of liquid placebo (or test) product and carefully rub it over all hand surfaces. After the liquid soap is spread, they then lather for 30 seconds. Subjects will wash their hands (to the wrists only) and fingernail areas by working the lather over them for 30 seconds and then rinse their hands in the bag up to the wrists only for 30 seconds.

The 1.0 liter volume of rinsate will be mixed well. A 5.0 ml aliquot will be removed from the bag with a sterile pipette and diluted in 0.1 M Sterile Stripping Suspending Fluid containing appropriate product neutralizers. The diluted aliquots will be pour-plated with Soybean-casein Digest agar plus neutralizers and incubated at $30° \pm 2°C$ for up to forty-eight (48) hours.

10. Laboratory Sink Method

Before beginning the wash, subjects will remove all jewelry from the hands and forearms and moisten their hands under running tap water ($40° \pm 2°C$). Subjects will dispense 2 pumps of liquid placebo (or test) product and carefully rub it over all hand surfaces.

After the liquid soap is spread, subjects will wash their hands, the fingernail area, and two-thirds of the forearms by working up a lather over them for 30 seconds. Subjects will rinse the hands and two-thirds of the forearms in running water for 30 seconds ($40° \pm 2°C$).

11. Incubation

All plates will be incubated for up to 48 hours at $30° \pm 2°C$. After incubation, the colonies on those plates that are countable will be enumerated.

12. Methods of Analysis

The plate count data collected from the study will be evaluated using the Mini-Tab® (or other) statistical computer software. The estimated $\log_{10}$ number of viable microorganisms recovered from both hand samples will be designated the "R-value." It is the adjusted average $\log_{10}$ colony-count measurement for each subject at each sampling time. Each R-value will be determined using the following formula:

$$R = \log_{10} [1000 \times C_i \times 10^{-D}]$$

where:

1000	= the number of mls of rinsate solution instilled into each zip-lock storage freezer bag ("Sterile Basin").
C_i	= the arithmetic average colony count of the 3 plate counts for each subject at a particular dilution level.
D	= the dilution factor of the rinsate.

13. Calculation of Microbial Reductions

Microbial counts (microbial populations recovered from both hands) per liter from the first and fifth basin for each subject on each sampling day are converted to $\log_{10}$ scale in order to linearize data, a requirement of the statistical models to be used. Data from both the first and fifth Sterile Basin Wash Technique samples for the 3 baseline days will be averaged for each individual. Geometric means of the data will be calculated as:

$$\text{Geometric Mean} = \text{antilog}_{10} \frac{1}{n} \sum \log_{10} x = (x_1 \cdot x_2 \cdots x_n)^{1/n}$$

In this case, because the $\log_{10}$ scale will have already been achieved from formula 1, formula 2 above will be modified as represented in formula 3.

$$\text{Geometric Mean} = \text{antilog}_{10} \frac{1}{n} \sum R$$

The following percent reductions will be calculated for both the test and baseline samples (using 50 test product subjects) collected after washes 1 and 5.

A = Geometric Mean of 1st Basin Wash Counts on Test Day (Day 12)

B = Geometric Mean of 5th Basin Wash Counts on Test Day (Day 12)

C = Geometric Mean of Three Baseline Basin Wash Counts for 1st wash

D = Geometric Mean of Three Baseline Basin Wash Counts for 5th wash

Percent Microbial Reduction Day 12, First Basin Wash Compared to the Pooled Baseline Samples from first washes:

$$First\ wash = \left\{ \frac{C - A}{C} \right\} \times 100$$

Percent Microbial Reduction Day 12, Fifth Basin Wash Compared to the Pooled Baseline Samples from the fifth washes:

$$Fifth\ wash = \left\{ \frac{D - B}{D} \right\} \times 100$$

14. Statistical Analysis

Exploratory Data Analysis (EDA), including Stem-Leaf Ordering, Letter Value Plots, and Box Plots, will be generated to assure that the data (R-values of the microbial counts) will approximate a Normal (Gaussian) Distribution.[81] Any outliers will be noted. If the distribution is normal, a series of Student's *t*-tests will be conducted and confidence intervals generated for each of the sample periods using the 0.05 level of significance for Type I (α) error. If the data do not approximate the normal distribution, a Mann-Whitney U test, the nonparametric analog of the Student's *t*-test, will be used in the analysis.

A pre-post experimental design will be used to evaluate the antimicrobial effectiveness of the extended antimicrobial product use period.

	Preproduct application			Postproduct application
R	$O(1)_{BL1}$	$O(5)_{BL1}$	$O(8)_{BL1}$	A	$O(12)_{T1}$
R	$O(1)_{BL5}$	$O(5)_{BL5}$	$O(8)_{BL5}$	A	$O(12)_{T5}$

where:

R   = Human Subjects Randomly Assigned to the Study

A   = Independent Variable (test antimicrobial soap)

$O(I)$ = Dependent Variable (microbial hand counts), Day I, for Baseline (BL) or Test (T) for i^{th} wash

IVa. GENERAL-USE HANDWASH EVALUATION (see Section VI.   Foodhandler Antimicrobial Handwash Evaluation)

V. EVALUATING ANTIMICROBIAL BODY SOAPS/LOTIONS

There are two general procedures for evaluating these products: (1) the axilla cup scrub methods and (2) the full body shower wash.

A. Axilla Cup Scrub Method

For evaluating body soaps and lotions, the most common test in the industry is the Axilla Cup Scrub Method. The objective of this evaluation is to determine the ability of the product to keep the microbial populations below a baseline number over periods of both 12 and 24 hours after a single-use application of the antimicrobial product. The underarm is used as the sample site because of the relatively large number of microorganisms which naturally reside there.

Both underarm regions are used. A single product can be applied to both sites or 1 product at the left side and a second at the right. Only the underarms are washed with the product for this procedure. Generally, the protocol does not require wearing sterile gauze at the evaluation sites to prevent cross-contamination. If the product requires more than a single application to be effective or acquires greater efficacy with prolonged and continuous use, this evaluation may not be applicable.

B. Full Body Shower Wash Evaluation

A full body shower/bath wash evaluation is a modified version of the full body presurgical wash evaluation. The axilla and/or the inguinal regions are used for testing in this evaluation. Subjects have baseline samples collected at the test sites on baseline days 1 and 5.[15]

On the first day of the test week, subjects use the product to shower in water at a controlled temperature for a specified length of time. They are immediately sampled by cup scrub at each test site upon completion of the shower. The test sample site is then protected with a sterile, nonocclusive bandage to assure no cross-contamination occurs.

Cup scrub samples are then collected at 12 and 24 hours postshower. Subjects continue to shower using the product for an additional 4 days. On the fifth day of the test, subjects are again sampled immediately after the shower, as well as 12 and 24 hours postshower.

Using this evaluation, an accurate and reliable baseline population can be determined; an immediate postshower measurement is provided in addition to the persistence evaluations of the Axilla Cup Scrub method. Also, the residual properties of the product can be determined by virtue of its repeated use for 5 consecutive days.

C. Sample Protocols

1. Purpose

The objective of this evaluation is to determine the efficacy of single applications of 2 bar soap test products in reducing the levels of microbial flora residing in the axillary regions of subjects. Test subjects will refrain from using antibacterial/antimicrobial soaps, medicated lotions and creams, underarm deodorants/antiperspirants, and dandruff shampoos throughout the course of the study. Subjects will not bathe or shower within 24 hours of the test product applications or laboratory sampling. Subjects will be provided for their use a kit containing personal hygiene products that do not compromise testing. The study consists of a 15 day washout period during which subjects use only products from the supplied kit and are sampled at the axillae on day 12, using the cup scrub method. On day 16, each test subject will utilize the 2 test products, one for each of their underarms. At 12 and 24 hours after product application, each subject will have each of their axillae sampled using the cup scrub procedure.

2. Scope

This test method is designed to determine the ability of 2 antimicrobial soap products to reduce the microbial flora residing in the axillary regions.

3. Test Material

The products to be evaluated are:

Test Product #1
Lot Number: _____
Expiration Date: _____

Test Product #2
Lot Number: _____
Expiration Date: _____

4. Test Methods

Subjects. A sufficient number of overtly healthy subjects over the age of 18, but under the age of 70, will be admitted into the study to ensure that 20 subjects complete testing. Subjects will be of mixed sex and age; all will be free from clinically evident dermatoses or injuries to the underarms. All subjects will be provided with study descriptions and will sign Informed Consent Forms prior to participating in the study.

Concurrent Treatment. No subject will be admitted into the study who is currently using topical or systemic antimicrobials, or any other medication known to affect the normal microbial flora of the skin.

5. Pretest Period

The 15 day period prior to the product-use portion of the study will be designated the "pretest" period. During this time, subjects will avoid the use of medicated soaps, lotions, shampoos, and deodorants, as well as avoid skin contact with solvents, acids, and bases. *Deodorants/antiperspirants must not be used in the underarm area by study participants throughout the course of this study.* Subjects will be instructed to use only the personal hygiene products provided to them while participating in the study. Subjects will also avoid using UV tanning beds and bathing in pools and/or hot tubs containing chlorine or other biocides. Subjects must not shave the axillary sites to be treated for five days prior to using the assigned test products. The subjects will not be allowed to bathe or shower within the 24 hour period prior to their sampling times. This regimen will allow for the stabilization of the normal microbial flora of the skin.

Before the initiation of this study, the Protocol study description will be given to subjects. Trained laboratory personnel will explain the study to each participant and will be available to answer any questions which may arise. Informed Consent Forms and any other supportive material relevant to the safety of the subjects will be supplied by principal investigators, following their review and approval by an Institutional Review Board (IRB). The primary purpose of the IRB is the protection of the rights and welfare of the subjects involved in the study (reference CFR 21, Parts 50 and 56). This study will begin only after IRB approval of all documents and procedures has been obtained.

Baseline samplings of the axillae will be performed on day 12 of the pretest period utilizing the cup scrub sampling technique.

6. Test Period

On day 16, subjects will be washed with one of the two test antimicrobial soaps at the right axilla first (which product at which site randomly assigned), and then with the second product at the left axilla, following the label instructions provided. T-shirts will be provided for subjects to wear for the duration of testing. The t-shirt must be worn at all times by the subject and must not be washed.

Twelve (12) and 24 hours after the products have been applied, both axillae will be sampled using the cup scrub sampling technique. The 12-hour sample will be taken from the lower portion of the axilla and the 24-hour sample will be taken from the upper portion of the axilla. The two sampling sites must not overlap.

7. Cup Scrub Sampling Technique

The cup scrub sampling technique is performed as follows:

Five (5) milliliters of Sterile Stripping Suspending Fluid, consisting of 0.1% Triton X-100 (Aldrich Chemical Co., Milwaukee, WI) in 0.1 ml/L phosphate-buffered saline solution (pH 7.8), will be added to the cylinder. The skin area inside the cylinder will then be massaged in a circumferential manner with a sterile rubber policeman for 2 minutes. One (1) ml aliquots of this solution are removed and plated in duplicate at appropriate dilutions on Soybean-casein Digest agar with neutralizers.

The plates will be incubated at $30° \pm 2°C$ for 48 hours to 72 hours. Those plates providing colony counts between 25 to 250 will be preferentially utilized in this study. If no plates provide counts in the 25 to 250 range, those plates having counts closest to that range will be used in determining the number of viable microorganisms. If 10^0 plates give an average count of zero, the average plate count will be expressed as 1.00. This is done for mathematical reasons. In $\log_{10}$ scale, the $\log_{10}$ of zero is undefined, but the $\log_{10}$ of 1.00 is zero.

8. Methods of Analysis

The plate count data collected from this study will be evaluated using the Mini-Tab® (or other) statistical computer software.

The estimated $\log_{10}$ number of viable microorganisms recovered from the axilla samples will be designated the "R-value." It is the adjusted average $\log_{10}$ colony count measurement for each subject at each sampling time. Each R-value will be determined using the following formula:

$$R = \log_{10}\left(\frac{5 \times C_i \times 10^{-D}}{A}\right)$$

where:

A = inside diameter of the cup scrub sampling cylinder.

5 = the number of milliliters of sterile stripping suspending fluid instilled.

C_i = the arithmetic average colony count of the duplicate plate counts for each subject at a particular dilution level.

D = the dilution factor of the rinsate.

Note: A $\log_{10}$ transformation will be performed on the collected data to convert it to a linear scale. A linear scale, more appropriately a $\log_{10}$ linear scale, is a basic requirement of the statistical models used in this study.

9. Statistical Analysis

The MiniTab® (or other) statistical computer package will be used for all statistical calculations. An Exploratory Data Analysis (EDA), including Stem-Leaf Ordering, Letter Value Plots, and Box Plots, will be generated to assure that the R-values approximate a normal distribution. If this is the case, a series of Student's *t*-tests will be conducted and confidence intervals determined for the sample periods using the 0.05 level of significance for Type I (α) error. If the data do not approximate the normal distribution, the non-parametric analog of the Student's *t*-test, the Mann-Whitney U test, will be used. Any outlier values obtained in the analysis will be noted. If a laboratory error is determined to be the cause, these data will not be used in the statistical evaluation; however, they will be included in the raw data.

The Students *t*-test or the Mann-Whitney U test will be adjusted for "multiple comparisons" using the procedure described by Dixon and Massey.[77] An experimental design will be utilized to evaluate and compare the antimicrobial effectiveness of the two products.

VI. FOODHANDLER ANTIMICROBIAL HANDWASH EVALUATION

A. Purpose

The objective of this test is to evaluate the antimicrobial efficacy of one handwash product formulation in reducing the level of transient microbial flora (contaminants) on the hands after a single, as well as multiple handwashes. Fifteen (15) human subjects will be used in this evaluation. The study will consist of a

7-day washout period followed by a 1-day treatment (product use) period. Enrolled subjects will be instructed to refrain from using antibacterial/antimicrobial soaps, medicated lotions and creams, as well as dandruff shampoos until the study has been completed. Subjects will be provided a personal hygiene kit containing products to be used which do not compromise testing. During the treatment period, subjects' hands will be contaminated with *Escherichia coli*, a generally benign bacterial species distinguishable from the other bacterial flora on the hands through use of differential culture media. Subjects' hands will be contaminated 11 consecutive times and sampled at baseline and 3 times over the course of 10 contamination/test product wash cycles. The first contamination/sample cycle will be to determine the baseline count using the glove juice sampling method, followed by a handwash with bland (non-antimicrobial) soap. The subjects' hands will be contaminated a second time, followed by a handwash with the test product and a glove juice sample. The second contamination and sample cycle will be conducted after the fifth product utilization wash (contamination cycle 6) and the final cycle after the tenth product utilization wash (contamination cycle 11).

1. Scope

This test method is designed to determine the ability of an antimicrobial product formulation to reduce transient microbial flora (contaminants) when used in a handwashing procedure utilizing the transient marker microorganism, *Escherichia coli*.

2. Test Material

The product to be evaluated is:

> Test Product
> Lot Number: _____
> Manufacture Date: _____

> Placebo Product
> Lot Number: _____
> Manufacture Date: _____

3. Test Methods

Subjects. A sufficient number of overtly healthy subjects at least the age of 18, but under the age of 70, will be admitted into the study to ensure that 15 subjects complete the study. Subjects will be of mixed sex and age. On the day of testing, all subjects will be free from clinically evident dermatoses or serious injuries to the hands and forearms. All subjects will be thoroughly instructed in the study design and will sign the Informed Consent Forms prior to participating in the study.

Concurrent Treatment. No subject will be admitted into the study who is currently using topical or systemic antimicrobials, or any other medication known to affect the normal microbial flora of the skin.

4. Pretest Period

The seven days prior to the test portion of the study will constitute the pre-test period. During this time, subjects will avoid the use of medicated soaps, lotions, deodorants, and shampoos, as well as skin contact with solvents, detergents, acids, and bases. Subjects will be supplied a personal hygiene kit containing a bland soap, shampoo, deodorant, and lotion, as well as a pair of rubber gloves. Subjects must use the contents of this kit exclusively for personal hygiene needs during their participation in the study. They must avoid contact with products on the restricted list. The rubber gloves must be worn when exposure to antimicrobials is unavoidable. Subjects will also avoid using UV tanning beds and bathing in biocidally-treated (e.g., chlorinated) pools and/or hot tubs. This regimen will allow for stabilization of the normal microbial populations residing on the hands.

Before the initiation of this study, the Protocol study description will be given to subjects. Informed Consent Forms and any other supportive material relevant to the safety of the subjects will be supplied by the principal investigators, following their review and approval by an Institutional Review Board (IRB). The primary purpose of the IRB is the protection of the rights and welfare of the subjects involved (reference CFR 21, Parts 50 and 56).

A study description and Informed Consent statements will be provided to each subject prior to their beginning the study. Trained laboratory personnel will explain the study to each participant and will be available to answer any questions that may arise.

Neutralization. A neutralization assay will be performed to assure that the neutralizers employed in this evaluation adequately inactivate the antimicrobial properties of the test product. If desired, a wash with a surgical scrub product may be added as well.

5. Test Period

A period of seven days will constitute the test week. Each subject will be employed 1 day of that week for about a 2-hour test period. A practice wash using the nonantimicrobial bland soap and the application procedure prescribed for the test product will precede the actual test portion of this study. The practice wash ensures that the subject understands the wash procedure.

After the practice wash has been completed, a 5.0 ml aliquot of a suspension containing approximately 1.0×10^8 cfu/ml of *Escherichia coli* (ATCC #11229), will be transferred into each subject's cupped hands. The inoculum will then be distributed evenly over both hands and up to the wrist, via gentle continuous massage for 45 seconds. After a 2-minute air-dry, the glove juice sampling procedure will be performed. This first inoculation cycle will provide the baseline inoculation levels. It will be followed by a 30-second nonmedicated soap handwash. The temperature of the water used for this and all subsequent wash cycles will be controlled at $40° \pm 2°C$.

The microbial inoculum will again be distributed over both hands up to the wrist, via gentle continuous massage for 45 seconds. After a 2-minute air-dry, the 15 subjects will wash with the test product according to the prescribed use directions. This will be followed by the glove juice sampling procedure.

This inoculation/wash procedure will be repeated a total of 11 consecutive times with a minimum of 5 and a maximum of 15 minutes between applications. Samplings of the hands for residual *Escherichia coli* will be performed after inoculation/product wash cycles 1, 5, and 10 using the glove juice sampling procedure.

6. Glove Juice Sampling Procedure

The glove juice sampling procedure will be performed as follows:

After the prescribed wash and rinse, the hands will be lightly dried with disposable paper towels and powder-free sterile gloves will be donned. At the designated sampling times, 75.0 ml of Sterile Stripping Suspending Fluid (SSSF) without product neutralizers will be instilled into the glove. The wrist will be secured, and an attendant will massage the hand through the glove in a standardized manner for 60 seconds. Aliquots of the glove juice (dilution 10^0) will be removed and serially diluted in 0.1% M Phosphate Buffered Solution with appropriate neutralizers.

Triplicate spread plates will be prepared from each of these dilutions, using MacConkey Agar with neutralizers. The *Escherichia coli* will produce purple colonies with a greenish metallic sheen on this differential medium, distinguishing it from the yellow colonies of the normal skin flora. The plates will be incubated at $30° \pm 2°C$ for approximately 48 hours. Those plates providing colony counts of purple colonies between 25 to 250 will be preferentially utilized in this study. If no plates provide counts in the 25 to 250 range, the plates closest to that range will be counted and used in determining the number of viable microorganisms. If 10^0 plates give an average count of zero, the average plate count will be expressed as 1.00. This is done for mathematical reasons. In $\log_{10}$ scale, the $\log_{10}$ of zero is undefined, but the $\log_{10}$ of 1.00 is zero. The number of viable bacteria recovered is obtainable from the formula $75 \times$ Dilution Factor $\times$ Mean Plate Count for the 3 plates.

Following the final product wash/sample cycle, the subjects will be required to wash with 70% isopropyl alcohol for 1 minute, air-dry the hands, and rinse under tap water in order to remove any remaining *Escherichia coli* from the hands.

7. Methods of Analysis

The plate count data collected from this study will be entered and evaluated using MiniTab® (or other) statistical computer software.

The estimated $\log_{10}$ number of viable microorganisms recovered from each hand will be designated the "R-value." It is the adjusted average $\log_{10}$ colony count measurement for each subject at each sampling time. Each R-value will be determined using the following formula:

$$R = \log_{10} [75 \times C_i \times 10^{-D}]$$

where:

75   = the amount of stripping solution instilled into each glove

C_i = the arithmetic average colony count of the triplicate plate counts for each subject at a particular dilution level

D   = the dilution factor

Note: The $\log_{10}$ transformation is performed on these data to convert them to a linear scale. A linear scale, more appropriately a $\log_{10}$ linear scale, is a requirement of the statistical models to be used.

8. Statistical Analysis

A pre/post-experimental design will be utilized to evaluate and compare the antimicrobial effectiveness of the test product:

Preproduct application	Postproduct application
R A O_{BL}	O_1 O_5 O_{10}

where:

R   = Subjects randomly assigned to the study

A   = Independent variable: Liquid test formulation

O_i = Dependent variable: Microbial counts at baseline (BL) and after product use washes 1, 5, and 10

Exploratory Data Analysis will be performed on the data. Stem-Leaf Ordering, Letter Value Plots, and Box Plots will be generated to assure the data collected approximate normal distribution. If this is the case, a series of Student's *t*-tests will be conducted using the 0.05 level of significance for Type I (α) error. If the data do not approximate a normal distribution, the nonparametric analog of the Student's *t*-test, the Mann-Whitney U test will be used. Any outlier values will be noted. If laboratory error is determined to be the cause, those data will not be used in the statistical evaluation, but they will be included among the raw data.

VII. MEDICAL/HEALTHCARE INDUSTRY ANTIMICROBIALS

In this area, the most frequently used topical antimicrobial studies are the healthcare personnel handwash, the surgical scrub, and the preoperative skin-prepping evaluations. These evaluations have received a great deal of input from the FDA and are presented in detail in the Tentative Final Monograph (CFR 21, Parts 333 and 369, Part III; 17 June, 1994). Examples of the basic working protocols follow.

A. Sample Protocols

1. Healthcare Personnel Handwash

Purpose. The purpose of this study is to examine the antimicrobial efficacy of 1 healthcare personnel handwash test product, 1 vehicle (test product without active ingredient), and 1 control product.

Scope. The antimicrobial effectiveness of 1 test healthcare personnel handwash formulation, 1 vehicle, and 1 control product will be determined utilizing 30 human subjects per formulation (a total of 90 subjects) over the course of 10 handwashes, with microbial samples taken after washes 1, 3, 7, and 10.

2. Test Material

The products to be evaluated are:

Test Product:

Lot Number: _____
Manufacture Date: _____

Vehicle Product:

Lot Number: _____
Manufacture Date: _____

Positive Control Product:

Lot Number: _____
Manufacture Date: _____

3. Test Methods

Subjects. A sufficient number of overtly healthy subjects over the age of 18, but under the age of 70, will be admitted into the study to ensure that 90 subjects complete the study (30 subjects per product configuration). Subjects will be of mixed sex and age and free from clinically evident dermatoses, open wounds, injuries to the hands or forearms, hangnails, or other disorders which may compromise the subject. All subjects will be familiarized with the study protocol and will sign Informed Consent Forms prior to participating in the study.

Concurrent Treatment. No subject will be admitted into the study who is currently using topical or systemic antimicrobials, or any other medication known to affect the normal microbial flora of the skin.

4. Pretest Period

The seven days prior to the test portion of the study will comprise the pretest period. During this time, subjects will avoid the use of medicated soaps, lotions, deodorants, and shampoos as well as skin contact with solvents, detergents, acids, and bases. Subjects will be supplied a personal hygiene kit containing nonmedicated soap, shampoo, deodorant, lotion, and latex gloves to be worn when contact with restricted materials cannot be avoided. Subjects must use the contents of this kit exclusively during their participation in the study. They must avoid contact with products on the list of restricted materials. Subjects will also avoid using UV tanning beds and bathing in chlorinated pools or hot tubs. This regimen will allow for stabilization of the normal microbial populations residing on the hands.

Before the initiation of this study, the Protocol study description will be given to subjects. Trained laboratory personnel will explain the study to each participant and will be available to answer any questions which may arise. Informed Consent Forms and any other supportive material relevant to the safety of the subjects will be supplied by principal investigators, following their review and approval by an Institutional Review Board (IRB). The primary purpose of the IRB is the protection of the rights and welfare of the subjects involved in the study (reference CFR 21, Part 56). This study will begin only after IRB approval of all materials and procedures has been obtained.

Neutralization. A neutralization assay will be performed to assure that the neutralizers employed effectively neutralize the antimicrobial activity of the product in this evaluation.

5. Experimental Period

The test will take place over a 14-day period. Each subject will be employed only 1 day of that period for a 4 to 5 hour period. A 30 second practice wash will be performed using a non-antimicrobial ("bland") soap and the same application procedure intended for that subject's test product. The practice wash will ensure that the subject understands the wash procedure. The temperature of the water used for this and all subsequent wash cycles will be controlled at $40° \pm 2°C$.

On the designated test day, a 5.0 ml aliquot of a suspension containing at least 1.0×10^8/ml *Serratia marcescens* (ATCC# 14756, red-pigmented strain) will be transferred into each subject's cupped hands. The inoculum will then be distributed evenly over both hands and not reaching above the wrists, via gentle continuous massage for 45 seconds. After a 2-minute air-dry, the glove juice sampling procedure will be performed. This inoculation/sampling cycle will provide the baseline inoculation levels. It will be followed with a 30-second bland soap handwash.

The microbial inoculum will again be distributed evenly over both hands, and not reaching above the wrists, via gentle continuous massage for 45 seconds. After a 2-minute air-dry, the subjects will wash with their assigned product configuration as specified by the sponsor-supplied use directions. This test wash will be followed by the glove juice sampling procedure.

This inoculation/product wash procedure will be repeated 10 consecutive times, with a minimum of 5 and a maximum of 15 minutes between applications. Samplings of the hands for remaining *S. marcescens* will be performed after inoculation/product wash cycles 1, 3, 7, and 10, using the glove juice sampling procedure.

6. Glove Juice Sampling Procedure

The glove juice sampling procedure will be performed as follows:

Following the prescribed wash and rinse, excess water will be shaken from the hands and powder-free, loose-fitting sterile latex gloves will be put on. At the designated sampling time, 75 ml of Sterile Stripping Suspending Fluid

(SSSF) without product neutralizers will be instilled into the glove. The wrist will be secured and an attendant will massage the hand through the glove in a uniform manner for 60 seconds. Aliquots of the glove juice (dilution 10^0) will be removed and serially diluted in Butterfield's Buffered Phosphate Diluent (BBP) solution with appropriate product neutralizers.

Spread plates will be prepared in triplicate from each of these dilutions using Soybean-casein Digest agar and incubated at $25° ± 2°C$ for approximately 48 hours. Those plates providing colony counts between 25 to 250 will be preferentially utilized in this study. If no plates provide counts in the 25 to 250 range, those plates with counts closest to that range will be used in determining the number of viable microorganisms. If 10^0 plates give an average count of zero, the average plate count will be expressed as 1.00. This is done for mathematical reasons. In $\log_{10}$ scale, the $\log_{10}$ of 0 is undefined, but the $\log_{10}$ of 1.00 is 0. The number of viable microorganisms recovered is obtainable from the formula:

$75 \times$ Dilution Factor $\times$ Mean Plate Count for the 3 plates.

Following the final product application cycle (application #10), the subjects will be required to wash with 70% isopropyl alcohol for 1 minute, air-dry their hands, and rinse under tap water in order to remove any remaining *Serratia marcescens* from the hands. If desired, a wash with a surgical scrub product may be added as well.

7. Methods of Analysis

The plate-count data collected from this study will be evaluated using MiniTab® (or other) statistical computer software. The estimated $\log_{10}$ number of viable microorganisms recovered from each hand will be designated the "R-value." It is the adjusted average $\log_{10}$ colony count measurement for each subject at each sampling time. Each R-value will be determined using the following formula:

$R = \log_{10} [75 \times C_i \times 10^{-D}]$

where:

75 = the amount of stripping solution instilled into each glove.

C_i = the arithmetic average colony count of the three plate counts for each subject at a particular dilution level.

D = the dilution factor.

Note: A $\log_{10}$ transformation will be performed on these data to convert them to a linear scale. A linear scale, more appropriately a $\log_{10}$ linear scale, is a requirement of the statistical models to be used.

8. Statistical Analysis

A pre-post experimental design will be utilized to evaluate and compare the antimicrobial effectiveness of the test product.

Preproduct application	Postproduct application
R A(1) $O(1)_{BL}$	$O(1)_1$ $O(1)_3$ $O(1)_7$ $O(1)_{10}$
R A(2) $O(2)_{BL}$	$O(2)_1$ $O(2)_3$ $O(2)_7$ $O(2)_{10}$
R A(3) $O(3)_{BL}$	$O(3)_1$ $O(3)_3$ $O(3)_7$ $O(3)_{10}$

where:

R   = subjects randomly assigned to 1 of 3 products in this study.

A(I)   = independent variables: the product configuration "I" assigned to each subject, where I = 1 for Test Product, 2 for Vehicle, and 3 for Control Product.

$O(1)_I$ = dependent variables: Microbial Counts at Baseline (BL) and after the ith product use (Washes 1, 3, 7, and 10)

Prior to performing a statistical analysis, exploratory Data Analysis will be performed on the data. Stem-Leaf Ordering, Letter Value displays, and Box Plots will be generated to assure the data collected approximate the Normal (Gaussian) Distribution.[81] If this is the case, a series of Student's t-tests will be conducted using the 0.05 level of significance for Type I (α) error. If the data do not approximate a normal distribution, the nonparametric analog of the Student's t-test, the Mann-Whitney U test will be used. Any outlier values will be noted. If laboratory error is determined to be the cause, those data will not be used in the statistical evaluation, but they will be included among the raw data.

B. Surgical Scrub

1. Purpose

The purpose of this study is to evaluate and compare the antimicrobial efficacy of three surgical scrub product configurations.

2. Scope

This study will evaluate and compare the antimicrobial efficacy of three surgical scrub product configurations. Antimicrobial efficacy will be measured in terms of three parameters—the immediate, the persistent, and the residual antimicrobial effects. At least 18 human subjects will be employed for each of the three product configurations for a total of 54 subjects.

3. Test Materials

Products to be used in this study are:

Test Product:

Lot Number: _____
Expiration Date: _____

Vehicle Product:

Lot Number: _____
Expiration Date: _____

Reference Product:

Lot Number: _____
Expiration Date: _____

4. Test Methods

Subjects. A sufficient number of overtly healthy subjects over the age of 18, but under the age of 70, will be admitted into the study to ensure that 54 subjects complete the study (18 subjects per product configuration). Subjects will be of mixed sex and age and free of clinically-evident dermatoses or injuries to the hands or forearms. All subjects will be familiarized with the study protocol and will sign Informed Consent Forms prior to participating in the study.

Concurrent Treatment. No subject will be admitted into the study who is currently using topical or systemic antimicrobials, or any other medication known to affect the normal microbial flora of the skin.

5. Pretest Period

The 14 days prior to the baseline portion of the study will constitute the pretest period. During this time, subjects will avoid the use of medicated soaps, lotions, deodorants, and shampoos, as well as skin contact with solvents, detergents, acids, and bases. Throughout the course of the study, subjects are to use only the nonantimicrobial personal products which are supplied to them. This regimen will allow for stabilization of the normal microbial populations residing on the hands.

6. Neutralization

A neutralization study will be performed prior to beginning actual testing in order to assure that the neutralizer(s) used effectively neutralize the antimicrobial properties of the products.

7. Baseline Period

Baseline measurements will be conducted over a week's time (baseline week) on days 1, 3, and 5, using a non-antimicrobial (bland) soap. The baseline count of the resident microbial populations will be used to evaluate eligibility for the study, as well as establish baseline values for each subject. Only those subjects with baseline counts of at least 1.5×10^5 organisms per hand will be selected to continue the study.

The baseline determinations will utilize the same sampling and recovery techniques used for the experimental period. Both hands will be sampled in baseline determinations. Subjects will not wash their hands for at least 2 hours prior to baseline samplings. Subjects will clean under fingernails with a nail stick and clip fingernails to a free-edge of less than or equal to 2 mm. All jewelry must be removed from the hands and arms.

Baseline determinations will be conducted as follows: The hands and the forearms up to 2/3 the distance from wrist to elbow will be washed in at $40° \pm 2°C$ water for 30 seconds using a bland soap. Samples will then be taken immediately (within 1 minute of scrubbing) using the glove juice sampling procedure.

8. Experimental Period

Subjects accepted into the study will be randomly assigned to 1 of the 3 product groups and be sampled as outlined in Tables 1 and 2. The test products will be used according to the directions supplied by the sponsor. The water temperature used for the surgical hand scrub will be regulated at $40° \pm 2°C$. The subjects will wash with the assigned products once on test days 1 and 5 and 3 times on test days 2, 3, and 4. The hands will be sampled after the first scrub procedure

Table 1 Surgical Scrub Schema

Test day	1	2	3	4	5
Number of scrubs per day	1	3	3	3	1

Left and right hands will be randomly sampled at times indicated in the sampling schedule using the glove juice sampling procedure.

Table 2 Sampling Schedule: Number of Hands Sampled at Indicated Time After Treatment

		Hour		
	Day	0	3	6
Baseline week: baseline	1, 3, and 5	108	0	0
Test week: test product 2% chlorhexidine gluconate	1, 2, and 5	12	12	12
Test week: vehicle product test product w/o the 2% chlorhexidine gluconate	1, 2, and 5	12	12	12
Test week: reference product 4% chlorhexidine gluconate	1, 2, and 5	12	12	12

The surgical scrub schema and the sampling schedule allow for the assessment of the immediate, persistent and residual effectiveness of the test products in reducing the normal microbial flora of the hands.

on days 1, 2, and 5. On each of the sampling days, the subjects' hands will be sampled at time zero (immediately after the scrub), 3 hours post-scrub, and/or 6 hours post-scrub.

9. Glove Juice Sampling Procedure

Following the prescribed wash and rinse, excess water will be shaken from the hands and loose-fitting, powder-free sterile gloves donned. At the designated sampling time, 75 ml of Sterile Stripping Suspending Fluid (SSSF) with neutralizers (pH 7.8–7.9) will be instilled into the glove. The wrist will be secured, and an attendant will massage the hand through the glove in a standardized manner for 60 seconds, paying particular attention to the area under the nails. Aliquots of the glove juice (10^0) will be removed and serially diluted in Butterfield's Buffered Phosphate Diluent (BBP) with neutralizers. Pour plates of Soybean-casein Digest agar with neutralizers will be prepared in duplicate with 1.0 ml aliquots of these dilutions. The solidified plates will be incubated at $30° \pm 2°C$ for approximately 48 hours. Those plates providing colony counts between 25 and 250 will be preferentially utilized in this study. If no plates provide counts in the 25 to 250 range, the plates having counts closest to that range will be used in determining the number of viable microorganisms. If 10^0 plates give an average count of zero, the population will be expressed as <10 rather than <25. The number of viable bacteria recovered is obtainable from the formula:

$$75 \times \text{dilution factor} \times \text{mean plate count.}$$

10. Statistical Analysis

A parametric, statistical model will be utilized to determine and compare the antimicrobial effectiveness of the test products. The 0.05 level of significance will be used in this study.

C. Preoperative Skin-Prepping Solutions

1. Purpose

The purpose of this study is to evaluate and compare the antimicrobial effectiveness of one preoperative skin-prepping formulation and one control product.

2. Scope

Thirty (30) human subjects will be employed, 15 subjects per each of the products, utilizing bilateral product applications. Two (2) separate anatomical sites, the inguinae and the abdomen, will be used in evaluating the products' immediate antimicrobial effects at 10 minutes and their persistent antimicrobial effects at 30 minutes and 6 hours post skin-prepping.

3. Test Material

The following test materials will be used in this evaluation:

Test Products

Product #1: (test formulation)

Lot Number: _____
Expiration Date: _____
Manufacture Date: _____

Product #2: (reference formulation)

Lot Number: _____
Expiration Date: _____
Manufacture Date: _____

Subjects. A sufficient number of overtly healthy subjects of at least the age of 18, but under the age of 70, will be admitted in the study to ensure that at least 30 subjects (15 per product configuration) complete the evaluation. Insofar as possible, subjects will be of mixed age, sex, and race, and all will be free of dermatoses or injuries to the skin areas being sampled.

Concurrent Treatment. No subject will be admitted into this study who is currently using a known topical or systemic antimicrobial, or any other medication known to affect the normal microbial flora of the skin.

4. Pretest Period

The 7-day period prior to the product use (test) portion of the study will be designated the "pretest" period. Subjects will be provided a personal hygiene kit that must be used for all personal care throughout the course of the study. Subjects will use only the assigned personal care soaps, lotions, shampoos, and deodorants, as well as avoid skin contact with solvents, acids, and bases. In addition, bathing in pools or hot tubs containing chlorine or other biocides will be prohibited. Subjects must not shave the anatomical sites proposed for treatment within the 5 days prior to being prepped. The subjects will not be allowed to bathe or shower within the 24 hour period prior to their sampling times. This regimen will allow for the stabilization of the normal microbial flora of the skin.

Before the initiation of this study, the protocol, study description, the subject information sheets, informed consent forms, and any other materials relevant to the subjects' safety will be reviewed and approved by an Institutional Review Board (IRB). The primary purpose of the IRB is to protect the rights and welfare of the subjects involved.

A description of this study and an informed consent statement (patient information) will be provided to subjects prior to their beginning the study. Trained laboratory personnel will explain the study to each participant and will be available to answer any questions that may arise.

5. Neutralization

A neutralization study will be performed prior to beginning actual testing in order to assure that the neutralizer(s) used effectively neutralize the antimicrobial properties of the products.

6. Baseline Week

Baseline sampling bilaterally in the inguinal area and on the abdomen in the vicinity of and on both sides of the umbilicus will be performed on two separate days of the baseline week, utilizing the cylinder sampling technique. The first baseline samples will be collected on day 1 of the baseline week and the second on day 4. In order for a particular subject to be accepted into the study, baseline colony counts at the inguinae must be at least 1.0×10^5 microbial organisms per cm^2. Baseline colony counts on the abdomen must be at least $1.0 \times 10^3/cm^2$.

7. Experimental Period

Subjects accepted into the study will be randomly assigned one of the two products. Prior to testing, the subjects will be questioned regarding adherence to the protocol and physically examined to ensure no evidence of injury or dermatosis is present at the sampling sites.

Subjects will then be sampled for baseline counts bilaterally at both the inguinae and the abdomen in the vicinity of the umbilicus. The inguinal and abdominal sites will be sampled 2 times for baseline. Subjects will then be prepped with the assigned product at the inguinae and the abdomen near the umbilicus following the label instructions provided.

A sample of each prepped area will be taken at 10 minutes post-prepping using the cylinder sampling technique. After this sample, a gauze and fenestration bandage (Smith & Nephews, Opsite I.V.) will be placed over the prepped areas to prevent any microbial contamination. Additional samples will be taken at 30 minutes and 6 hours postskin-prepping.

8. Cylinder Sampling Technique

The cylinder sampling technique will be performed as follows:

At the designated sampling times, a sterile cylinder will be held firmly onto the skin surface of the test site to be sampled. 2.5 ml of Sterile 0.1 M Phosphate Buffered Stripping Solution with 0.1% Triton X-100 and product neutralizers will be instilled into the cylinder, and the skin area inside the cylinder will be massaged in a circumferential manner for 1 minute with a sterile rubber policeman. The entire procedure will be repeated without lifting the cylinder to obtain a second aliquot, and the two aliquots of 2.5 ml each will then be pooled.

One (1) ml aliquots of the microorganism suspension (10^0 dilution) will be removed, plated and/or serially diluted (as appropriate) with Butterfield's Buffered Phosphate (BBP) Solution. Pour plates will be prepared in triplicate with 1 ml from each of these dilutions using Soybean-casein Digest agar and the plates will be incubated at $30° \pm 2°C$ for 72 ± 2 hours. Those dilutions yielding 25 to 250 colonies per plate will be preferentially counted. In cases where no plates provide counts in the 25–250 range, those plates with the highest number of countable colonies closest to that range or those from the lowest dilution will be used.

9. Data-Handling and Statistics

The estimated $\log_{10}$ number of viable microorganisms per cm^2 recovered from each sample site will be designated the "R-value." To convert the volumetric quantities collected during sampling into the number of colony-forming units per square centimeter (cm^2), the following formula will be employed:

$$R = \log_{10}\left[\frac{F\left(\dfrac{\Sigma c_i}{n}\right)10^{-D}}{A}\right]$$

where:

R   = The average colony count in $\log_{10}$ scale per cm^2 of sampling surface.

F   = total number of ml of stripping fluid added to the sampling cylinder. In this study, F is equal to 5 ml.

$\dfrac{\Sigma c_i}{n}$ = average of the triplicate plate counts used.

D   = dilution factor of the plate counts.

A   = inside area of the cylinder in cm^2.

Note: A $\log_{10}$ transformation will be performed on the collected data is to convert them to a linear scale. A linear scale, more appropriately a $\log_{10}$ linear scale, is a basic requirement of the statistical models used in this study.

10. Statistical Analysis

The MiniTab® (or other) statistical computer package will be used for all statistical calculations. An Exploratory Data Analysis (EDA), including Stem-Leaf Ordering, Letter Value Plots, and Box Plots, will be generated to assure that the R-values approximate a Normal (Gaussian) Distribution.[81] If this is the case, a series of Student's *t*-tests will be conducted and confidence intervals determined for the sample periods using the 0.05 level of significance for Type I (α) error. If the data do not approximate the Normal Distribution, the nonparametric analog of the Student's *t*-test, a Mann-Whitney U test, will be used.

Any outlier values obtained in the analysis will be noted. If laboratory error is determined to be the cause, these data will not be used in the statistical evaluation; however, they will be included in the raw data. The Student's *t*-test or the Mann-Whitney U test will be adjusted for "multiple comparisons" using the procedure described by Dixon and Massey.[77]

A pre-post experimental design will be utilized to evaluate and compare the antimicrobial effectiveness of the products.

Pre-product application			Post-product application		
R	A	$O_{A,BL,1}$ $O_{A,BL,2}$	$O_{A,T,10min}$	$O_{A,T,30min}$	$O_{A,T,6hr}$
R	C	$O_{C,BL,1}$ $O_{C,BL,2}$	$O_{C,T,10min}$	$O_{C,T,30min}$	$O_{C,T,6hr}$

where:

R = Subjects randomly assigned to each of the 2 groups. A random number series will be generated for this purpose.

A = Independent variable: In this study, A represents the test product.

C = Independent variable: In this study, C represents the reference product.

O_{xij} = Dependent variable: For test or control product (A or C), microbial counts per cm^2 at the ith period and the jth time.

x = Test (T) or Control (c) product.

I = Baseline (BL) or Test (T) period.

j = Period 1 or 2 for the baseline phase and times of 10 minutes, 30 minutes or 6 hours for test phase.

D. Indwelling Catheter/Preinjection Preparation Study

1. Purpose

The purpose of this study is to evaluate and compare the antimicrobial effectiveness of 1 antimicrobial solution in 2 different application packages, the vehicle in 2 different application packages, and 1 reference product.

2. Scope

This evaluation will measure the antimicrobial effectiveness of 1 test product presented in 2 different application packages. One (1) application package is used for indwelling catheter site preparations and the other is used for preinjection site preparations. The vehicle, presented in packages identical to those of the test product, and a reference product are to be used as controls for both the indwelling catheter and the preinjection preparation. A minimum of 45 human subjects (a minimum of 30 sites per formulation) will be employed, and product applications will be bilateral. The immediate and persistent antimicrobial efficacy of the formulations will be evaluated at three separate anatomical sites over a 24-hour period to support the claim of an indwelling catheter preparation. For the claim of a preinjection preparation, only the immediate antimicrobial efficacy will be evaluated at two separate anatomical sites.

3. Test Material

The products to be used in this evaluation are:

Test Product Applicator 1

Lot Number: _____
Expiration Date: _____

Test Product Applicator 2

Lot Number: _____
Expiration Date: _____

Product Vehicle Applicator 1

Lot Number: _____
Expiration Date: _____

Product Vehicle Applicator 2

Lot Number: _____
Expiration Date: _____

Reference Product

Lot Number: _____
Expiration Date: _____

4. Test Sites

Test product in Application Package 1 will be utilized only at the (a) deltoid and (b) gluteus maximus sites. Test product in Application Package 2 will be utilized only at the (a) subclavian, (b) femoral (inguinal), and (c) median cubital sites. The Vehicle and Reference product will be used at all anatomical sites.

Each subject will be randomly assigned 1 product configuration for bilateral product application. There are three possible product configurations. Both test product applications (Test product application 1 utilized at the deltoid and gluteus maximus sites and Test product application 2 utilized at the subclavian, femoral and median cubital sites) and the vehicle configurations; both Test product application and the reference; and the reference and Vehicle product application.

5. Test Methods

Institutional Review Board (IRB).   Prior to the initiation of this study, an Institutional Review Board (IRB) will be assembled to review and approve the Protocol, including the study description, informed consent forms, and any other supportive material relevant to the safety of the human subjects involved in the study (reference CFR 21, Parts 50 and 56). This study will begin only after IRB approval has been obtained for all facets.

Subjects. A sufficient number of overtly healthy subjects of at least the age of 18, but under the age of 70, will be admitted to the study to ensure that at least 45 subjects utilizing bilateral product application, complete the evaluation. Insofar as possible, subjects will be of mixed age, sex, and race, and all will be free of dermatoses or injuries to the sites being sampled.

Since both test product applicators will be used at separate sites on one side of each subject (test product applicator 1 at the deltoid and gluteus maximus; test product applicator 2 at the subclavian, femoral, and median cubital regions) each subject will be assigned 1 of 3 possible product configurations: (1) test products and vehicle; (2) test products and reference; and (3) reference and vehicle. Hence, each of the 45 subjects will receive 1 of these test configurations.

Trained laboratory personnel will review the Study Description with each subject. Upon completion of their review, each subject will be given a copy of the Study Description and will sign and date the Informed Consent Form and complete the Subject Confidential Information and Acceptance Criteria. Trained laboratory personnel will assign a subject number to each subject and complete the Informed Consent for each subject.

Concurrent Treatment. No subject will be admitted into this study who is currently using a known topical or systemic antimicrobial, or any other substance known to affect the normal microbial flora of the skin.

6. Pretest Period

The 14-day period prior to the product-use portion of the study will be designated the "pretest" period. During this time, subjects will use only the personal hygiene soaps, lotions, shampoos, and deodorants that are provided and will avoid skin contact with solvents, acids, and bases. In addition, subjects will avoid use of tanning booths and bathing in pools or hot tubs containing chlorine or other biocides. The subjects will not be allowed to bathe or shower during the 24-hour period prior to being sampled. This regimen will allow for the stabilization of the normal microbial flora of the skin.

7. Neutralization

Prior to initiation of this study, the adequacy of the neutralizers will be confirmed in accordance with Standard Practices for Evaluating Inactivators of Antimicrobial Agents Used in Disinfectant, Sanitizer, Antiseptic, or Preserved Products (ASTM E1054-91).[90]

8. Baseline Week

The week following the 14-day pre-test period will constitute the baseline week. Subjects will not bathe or shower within 24 hours of being sampled. All subjects will be sampled on days 1 and 4 of that week at the 5 selected anatomical sites per bilateral configuration. There will be a minimum of 72 hours between the time the baseline period ends and the experimental period begins. Based upon adequate baseline microbial counts from these 2 baseline assessments, subjects will be selected to complete this study. Baseline acceptance criteria for deltoid, gluteus maximus, median cubital, and subclavian areas will be at least 5.0×10^2 cfu/cm^2. Acceptable counts for the femoral (inguinal) region will be at least 1.0×10^5 cfu/cm^2.

All sample sites requiring clipping will be clipped at least 48 hours prior to test day 1. Clipping will be completed to assure that the bandaging material (used during the experimental test period) remains securely affixed to the test site for the 24-hour testing period. Prior to being sampled, the subjects will be questioned regarding their adherence to protocol restrictions. Sampling sites will also be examined visually to ensure no evidence of injury or dermatosis is present.

A third and final baseline sample will be collected at each test site prior to prepping/testing on test day 1. Sampling will be performed near the (1) subclavian vein; (2) the median cubital vein of the forearm; (3) the femoral vein in the inguinal area; (4) the deltoid region of the arm; and (5) the gluteus maximus (buttocks) using the cylinder sampling technique.

9. Test Period

Subjects will be questioned regarding their adherence to Protocol restriction requirements and examined visually to ensure no evidence of injury or dermatosis is present at the anatomical sampling sites. Subjects will refrain from showering or bathing during the 24-hour period prior to being sampled.

Subjects will not shower or bathe while they are in the 24-hour experimental period. Subjects may take sponge baths, but are to avoid washing the bandaged sampling sites.

10. Indwelling Catheter Preparation

On day 1 of the test week, the regions near the subclavian vein, the femoral vein, and the median cubital vein will be sampled for the final (third) baseline measurement and then prepped according to Sponsor's instructions using the assigned product. Samples will be taken immediately post-prepping using the cylinder sampling technique. A sterile bandage (Smith and Nephew bandage material) will be placed over the prepped area to help prevent microbial contamination. Additional samples will be taken 24 hours post-prepping using the cylinder sampling technique.

11. Preinjection Preparation

On day 1 of the test week, the deltoid and gluteus maximus sites will be sampled for the final (third) baseline measurement and then prepped using the assigned product according to Sponsor's instructions. Samples will be taken immediately post-prepping using the cylinder sampling technique.

12. Cylinder Sampling Technique

The cylinder sampling technique will be performed as follows:

At the designated sampling times, a sterile cylinder will be held firmly onto the skin surface of the test site to be sampled. 2.5 ml of Sterile 0.1 M Phosphate Buffered Stripping Solution with 0.1% Triton X-100 and product neutralizers will be instilled into the cylinder, and the skin area inside the cylinder will be massaged in a circumferential manner for 1 minute with a sterile rubber policeman. The entire procedure will be repeated to obtain a second aliquot, and the 2 aliquots of 2.5 ml each will be pooled.

One (1) ml aliquots of the microorganism suspension (10^0 dilution) will be removed, plated, and/or serially diluted (as appropriate) with Butterfield's Buffered Phosphate (BBP) Solution. Triplicate pour plates will be prepared with 1 ml from each of these dilutions using Soybean-casein Digest agar.

The inoculated plates will be incubated at $30° \pm 2°C$ for 72 ± 2 hours. Those dilutions yielding 25 to 250 colonies per plate will be preferentially counted. In cases where no plates provide counts in the 25–250 range, those plates with the highest number of countable colonies closest to that range, or the lowest dilution, will be used.

13. Test Duration

Testing will continue over a 1- to 2-month period with 3 groups of 15 people staggered over that time.

14. Data-Handling and Statistics

The plate-count data will be recorded on raw data sheets. The raw plate-count data will be transformed into $\log_{10}$ values to linearize them, a requirement of the statistical models used. The statistical level of significance will be set at $\alpha = 0.05$ and all confidence intervals will be reported at the 95% confidence level ($1 - \alpha$ or $1 - 0.05 = 0.95$). A MiniTab® (or other) computer statistical package will be used for all statistical calculations.

In order to convert the volumetric measurements into the number of colony-forming units per square centimeter (cm^2), the following formula will be employed:

$$R = \log_{10}\left[\frac{F\left(\dfrac{\Sigma c_i}{n}\right)10^{-D}}{A}\right]$$

where:

R = the average colony count in $\log_{10}$ scale per cm^2 of sampling surface.

F = total number of ml of stripping fluid added to the sampling cylinder. For this study, F is equal to five (5) ml.

$\dfrac{\Sigma c_i}{n}$ = average of the triplicate plate counts used for each sample collected.

D = dilution factor of the plate counts.

A = inside area of the cylinder in cm^2.

A pre/post experimental design will be used to evaluate the antimicrobial properties, for the preinjection design (reference Table 3).

Table 4 presents the indwelling catheter preparation design.

A biostatistical evaluation will be performed on the data to assess the immediate and persistent antimicrobial effects of the 2 test products, the vehicle, and the reference product. Statistical summaries will be generated in both tabu-

Table 3 Preinjection Design Evaluated at the Deltoid and Gluteus Maximus Sites

Preproduct application				Postproduct application
R	T_1	O_{BL1}	O_{BL2}	O_{BL3}	O_{T1}
R	V	O_{BL1}	O_{BL2}	O_{BL3}	O_{T1}
					O_{T1}
R	C	O_{BL1}	O_{BL2}	O_{BL3}	O_{T1}

where:

R = Human subjects randomly assigned to products
T_1 = Independent variable (test product #1)
V = Independent variable (vehicle product)
C = Independent variable (reference product)
$O_{i,j}$ = Dependent variable (microbial counts/cm^2)
    $_i$ = BL if baseline, T if test
    $_j$ = 1, 2, 3 for baseline samples or 1 if test sample
    (immediate)

Table 4 Indwelling Catheter Prep Design Evaluated at the Subclavian, Femoral, and Median Cubital Sites

Pre-product application					Post-product application		
R	T_2	O_{BL1}	O_{BL2}	O_{BL3}	O_{T1}	O_{T2}	O_{T3}
R	V	O_{BL1}	O_{BL2}	O_{BL3}	O_{T1}	O_{T2}	O_{T3}
R	C	O_{BL1}	O_{BL2}	O_{BL3}	O_{T1}	O_{T2}	O_{T3}

where:

R = Human subjects randomly assigned to products
T_2 = Independent variable (test product #2)
V = Independent variable (vehicle product)
C = Independent variable (reference product)
$O_{i,j}$ = Dependent variable (microbial counts cm^2)
  i = BL if baseline, T if test
  j = 1, 2, 3 for Baseline Samples
    OR
  1, 2, 3 for Test Samples
  1 if immediate
  2 if 24 hours
  3 if 72 hours

lar and graphical form for the 3 test samples times for the 5 selected anatomical sites.

The statistical summaries will utilize pooled baseline values, after the data for the 3 baselines have been shown to be statistically equivalent. The pooling of data is the preferred procedure, as it minimizes the error. The statistical summary will discuss any significant variances in performance among the test formulations, vehicle, or reference product, if they should occur. The Student's *t*-test will be used to analyze the data, when possible, for ease of interpretation by reviewers with limited statistical knowledge.

The anatomical sites evaluated will be handled separately in this evaluation. The level of significance for type I or α error is set at 0.05. However, all statistical comparisons will also be given as "P" values of observing a "t" calculated value as extreme or more extreme than that value, given the null hypothesis is true. A multiple *t*-test correction factor will be used in computing multiple *t*-tests, as described by Dixon and Massey.[77]

Since all 45 subjects cannot be used in a single day, a one-factor Analysis of Variance (ANOVA) model will be used to determine if testing done on different days introduced significant ($P < 0.05$) variance in the antimicrobial effects of the products evaluated.

E. Full Body Presurgical Wash

Before surgical procedures are performed, it is standard that the proposed operative site be prepared with an effective antimicrobial to reduce the microbial populations residing on the skin and, hence, the potential for surgery-associated infection. Povidone iodine and chlorhexidine gluconate (CHG) have been the two antimicrobials most commonly selected for preoperative patient skin preparation through the years. In efficacy trials with human volunteers whose baseline counts exceeded 10^5 organisms/cm^2, both antimicrobial products usually demonstrated at least a 3 $\log_{10}$ reduction in resident skin flora within 10 minutes of exposure.

Because the normal resident microbial populations of the abdominal and thoracic regions average approximately 10^3 organisms/cm^2, the vast majority of these flora are removed during the skin preparation process. At such anatomic sites as the inguinal region, however, where the microbial counts average 5.5×10^5 organisms/cm^2, the preoperative patient preparation, alone, may not be adequate in preventing postsurgical infection. In these anatomic areas with particularly high microbial counts, there will likely be significant numbers of microorganisms remaining on the prepared skin site. These organisms may have potential to cause postoperative infection, especially in immunocompromised patients.

Healthy human subjects older than 18 years, but younger than 70 years, are recruited. Insofar as possible, subjects are of mixed age, sex, and ethnic background. All subjects are free of clinically evident dermatoses and injuries to the skin areas being sampled. No subject will be admitted to the study who is using topical or systemic antimicrobial agents, or any other medication known to affect the normal microbial populations of the skin.

Once the subjects are admitted to the study, a 14-day pretest period is observed. During this time, subjects avoid the use of medicated soaps, lotions, shampoos, deodorants, chlorinated water baths, and ultraviolet light tanning beds, as well as skin contact with solvents, acids, and bases. This regimen permits stabilization of the normal microbial flora populations on the skin.

A 7-day baseline period follows the pretest period. Baseline skin-sampling is conducted on days 1, 3, and 5 at both the abdominal and inguinal regions. A 5-day test period follows. Subjects use the test products, as per instructions.

On test day 1, immediately after the shower procedure using the test products, each subject dries his or her body with a supplied, sterile, soft, absorbent terry towel. Each is then sampled immediately (within 10 minutes of showering) using the cylinder sampling technique and again at 3 and 6 hours after the shower wash. Identical sampling procedures are repeated on test days 2 and 5. Abdominal samples are taken from the region extending approximately 1 inch to the left and right of the umbilicus. The inguinal region is sampled at the uppermost inner aspect of the thigh in the inguinal crease. A sterile gauze material secured with adhesive tape is used to protect the sampling sites from transient microorganism contamination between samplings. The gauze bandage is removed for the 3 and 6 hour skin-samplings and then reapplied. The subjects are not permitted to take additional showers or baths during the test period.

1. Cylinder Sampling Technique

The cylinder sampling technique will be performed as follows:

At the designated sampling times, a sterile cylinder will be held firmly onto the skin surface of the test site to be sampled. 2.5 ml of Sterile 0.1 M Phosphate Buffered Stripping Solution with 0.1% Triton X-100 and product neutralizers will be instilled into the cylinder, and the skin area inside the cylinder will be massaged in a circumferential manner for 1 minute with a sterile rubber policeman. The entire procedure will be repeated without lifting the cylinder to obtain a second aliquot, and the two aliquots of 2.5 ml each will then be pooled.

One (1) ml aliquots of the fluid microorganism suspension (10^0 dilution) will be removed, plated and/or serially diluted (as appropriate) with Butterfield's Buffered Phosphate (BBP) Solution. Pour plates will be prepared in triplicate with 1 ml from each of these dilutions using Soybean-casein Digest agar, and the plates will be incubated at $30° \pm 2°C$ for 48 to 72 hours. Those dilutions yielding 25 to 250 colonies per plate will be preferentially counted. In cases where no plates provide counts in the 25–250 range, those plates with the highest number of countable colonies closest to that range or those from the lowest dilution will be used.

2. Data-Handling and Statistics

The estimated $\log_{10}$ number of viable microorganisms per cm^2 recovered from each sample site will be designated the "R-value." To convert the volumetric quantities collected during sampling into the number of colony-forming units per square centimeter (cm^2), the following formula will be employed:

$$R = \log_{10}\left[\frac{F\left(\dfrac{\Sigma c_i}{n}\right)10^{-D}}{A}\right]$$

where:

R   = The average colony count in $\log_{10}$ scale per cm^2 of sampling surface.

F   = total number of ml of stripping fluid added to the sampling cylinder. In this study, F is equal to 5 ml.

$\dfrac{\Sigma c_i}{n}$ = average of the triplicate plate counts used.

D = dilution factor of the plate counts.

A = inside area of the cylinder in cm^2.

Note: A log_{10} transformation will be performed on the collected data is to convert them to a linear scale. A linear scale, more appropriately a log_{10} linear scale, is a basic requirement of the statistical models used in this study.

3. Statistical Analysis

The MiniTab® (or other) statistical computer package will be used for all statistical calculations. An Exploratory Data Analysis (EDA), including Stem-Leaf Ordering, Letter Value Plots, and Box Plots, will be generated to assure that the R-values approximate a Normal (Gaussian) Distribution.[81] If this is the case, a series of Student's t tests will be conducted and confidence intervals determined for the sample periods using the 0.05 level of significance for Type I (α) error. If the data do not approximate the Normal Distribution, the nonparametric analog of the Student's t-test, a Mann-Whitney U test, will be used.

Any outlier values obtained in the analysis will be noted. If laboratory error is determined to be the cause, these data will not be used in the statistical evaluation; however, they will be included in the raw data. The Student's t-test or the Mann-Whitney U test will be adjusted for "multiple comparisons" using the procedure described by Dixon and Massey.[77]

A pre-post experimental design will be utilized to evaluate and compare the antimicrobial effectiveness of the products.

Preproduct Application			Postproduct Application		
R	A	$O_{A,BL,1}$ $O_{A,BL,2}$	$O_{A,T,10min}$	$O_{A,T,30min}$	$O_{A,T,6hr}$
R	A	$O_{A,BL,3}$ $O_{A,BL,2}$	$O_{A,T,10min}$	$O_{A,T,3hr}$	$O_{A,T,6hr}$
R	C	$O_{C,BL,1}$ $O_{C,BL,2}$	$O_{C,T,10min}$	$O_{C,T,30min}$	$O_{C,T,6hr}$
R	A	$O_{C,BL,3}$ $O_{A,BL,2}$	$O_{A,T,10min}$	$O_{A,T,3hr}$	$O_{A,T,6hr}$

where:

R = Subjects randomly assigned to each of the 2 groups. A random number series will be generated for this purpose.

A = Independent variable: In this study, A represents the test product.

C = Independent variable: In this study, C represents the reference product.

O_{xij} = Dependent variable: For test or control product (A or C), microbial counts per cm^2 at the ith period and the jth time.

x   = Test (T) or control (C) product

I   = Baseline (BL) or test (T) period

j   = Period 1, 2, or 3 for the baseline phase and times of 10 minutes, 3 hours, or 6 hours for test phase.

8
Overview of Statistical Methods

As I have emphasized earlier, a properly designed statistical evaluation is one that systematically collects, organizes, analyzes, and draws valid conclusions about the product(s) being evaluated. Hence, when one designs a study, it is critical that the objectives be explicitly stated, the data expected be well understood, and the statistical model applied be appropriate to clearly explicating the significance of the data. In introduction to the detailed overview of statistical methods that follows, I will re-emphasize a few of the important considerations of study design discussed in different context in Chapters 5 and 6.

I. THE PURPOSE OF THE EVALUATION

A description of the purpose of the evaluation must be well-considered and concise, because all subsequent aspects of the study derive from it. If this is not accomplished, particularly for complex studies, the original objectives often become more obscure. Then one is faced with backtracking through evaluation records, trying to determine what the original objectives were and, worst of all, having to make misfit data fit.[78]

Once the purpose of the evaluation has been determined, assurance of validity in study design must be considered. Validity in experimental designs encompasses two areas: internal and external validity.

A. Internal Validity

Internal validity is research-design validity. In particular, it deals with the way the study is designed, how sample data are collected, and how the study is experimentally controlled, especially with respect to investigator bias. Generally, an investigator has a "vested interest" in realizing success in the area of their primary professional interest. This bias must be taken into account.[2]

Fortunately, internal validity can be assured by using proper experimental design procedures (e.g., randomizing, blocking, and blinding the study, as discussed in Chapter 6). While there are a number of aspects to internal validity, two of the most common are historical and instrumentation validity.[80]

Historical validity assures that no event occurs between sample time measurements which biases the study results.[91] An example of negation of historical validity happened when a meat-packing plant conducted a handwashing efficacy study. The investigator initially took baseline samples of the employees that were accurate and reliable. The investigator then assigned individuals a test product but, unknown to the investigator, 7 of the 10 participants began using an antimicrobial soap in their personal hygiene practices. They wanted to look good for the investigator. The test product was credited with providing effective degerming properties but, in reality, the effect was due largely due to the use of the antimicrobial soaps.

Instrumentation validity is achieved by assuring that no extraneous events occur which affect the measuring instruments used in the experiment.[78] For example, recently, in a poultry-processing plant, a quality-control microbiologist used different agar media lots in assessing a handwashing study. The media lots, made by different manufacturers, differed significantly in nutritional characteristics and, therefore, varied in their support of microbial growth, thus biasing the data. The microorganisms grew well on one lot of media, but not the other. This growth difference was erroneously attributed to the test products.

B. External Validity

External validity refers to the extent to which the results of a specific study can be generalized to the population-at-large (population validity) or to all general environmental conditions (environmental validity).[92] An example of failure in population validity occurred when an investigator only used females of Nordic descent in a handwash efficacy evaluation. These women tended naturally to have low microbial population counts on their hands, and the investigator concluded washing with merely soap and water was an adequate hygienic procedure for employees in all company processing plants. However, it soon became apparent that this regimen was not as effective as it had seemed. A major reason was that men, as well as individuals from different ethnic backgrounds, had higher populations of resident microorganisms on their hand surfaces. A mild, non-antimicrobial soap and water wash were not enough to cleanse the hands of those employees.

Failure in environmental validity is also exemplified in this particular study. Not only did the investigator utilize women of Nordic descent, but the study was conducted in Montana in the winter, where microbial populations on the skin of the hands are relatively lower because of the dry, cold air. Higher microbial populations were encountered in others of the company's processing plants in Georgia, Louisiana, and southern Texas. Again, a mild, nonantimicrobial soap and water wash were not up to the task of cleansing adequately the hands of employees working in those regions of the country.

Controls for assuring external validity cannot be built into the experimental design. The singular approach to maximizing it is to have the same study conducted independently at different geographic locations and/or with different subject populations. If consistent results are observed and the same conclusions are drawn by different investigators, the external validity of the study is likely satisfactory, and the data are generalizable.

C. Statistical Methods

The vast majority of quantitative research designs utilize statistical methods to analyze data. The exact statistical model to be used depends, in part, on the data distribution generated (normal, skewed, bimodal, exponential, binomial, or others). The use of exploratory data analysis (EDA) procedures[93] (as discussed in Chapter 6) can help the investigator select the appropriate statistical model and develop an intuitive "feel" for the data before the actual statistical analysis begins.

It is also important that the experimental data collected are of linear scale, a requisite of most statistical models. For example, if one is evaluating the antimicrobial properties of a hand cleanser, the experimental data are the microbial colony counts from sampling of the skin of the hands. A problem is that microbial inactivation rates are usually not linear, but rather, are exponential. Hence, the exponential microbial count data must be transformed to linear scale. Let me present a real-life scenario. Say the average baseline microbial counts are 1.0×10^6. Following a wash procedure with an iodine product, the microbial population levels are 1.0×10^4, and an hour later, they are 1.0×10^5. Evaluating these data in exponential scale poses statistical problems, for they are nonlinear. However, if the logarithmic values of these data are used, the data are transformed to $\log_{10}$ linear scale, the log transformed values—6, 4, and 5, respectively—can now be utilized in linear statistical models.

As an added reminder, it is also necessary to establish the levels of both alpha (α) and beta (β) error, such that the appropriate number of test items—the sample size—to be evaluated can be determined in terms of the desired statistical confidence level. Recall that α error (type I error) is committed when one rejects a true-null hypothesis, and β error (type II error) is committed by accepting a false-null hypothesis.[92] In other words, an α error occurs when one concludes based on analysis that there is a difference between handwashing products or methods when there really is not; a β error occurs when one concludes, conversely, that there is no difference between handwashing products or methods when there really is. The easiest way to control both α and β errors is to use more test subjects so that the possibility of both α and β errors is reduced.[77] Otherwise, merely adjusting the α error to a very small level will increase the probability of β error.

D. Parametric Statistics

Parametric statistics such as the Student's t-test, linear regression, analysis of variance (ANOVA), and analysis of covariance (ANCOVA) utilize parameters (mean, standard deviation, and the variance) in evaluating data.[93] The data collected are termed "interval" data (102.915, 1×10^{-5}, $7.23914 \ldots$). Interval data are data which can be ranked, as well as be subdivided into an infinite number of intervals. Usually, interval data relate to some standard physical measurement (e.g., height, weight, blood pressure, or numbers of deaths per capita). On the other hand, subjective perception of, for example, pregnancy, prestige, and social stress do not present a "natural" interval category, although a number of research designs may erroneously categorize them as interval data. Extreme caution must be exercised to avoid using quantitative statistical research designs (parametric statistical models) to measure such data. Let us begin our discussion of parametric statistics with an explanation of what is meant by level of measurement. It is important to know that the two lowest (least informative) data forms—nominal and ordinal—are not appropriate for parametric statistical analysis, but must be evaluated using nonparametric methods.

E. The Levels of Measurement

1. Nominal Measurements

This is the weakest level of measurement. Nominal measurement data are classified into mutually exclusive categories without specific numeric value.[93] Numbers are assigned to the data simply to distinguish or identify the individual items. For example, at a stock car race, a number is painted on the side of each car. The number chosen means nothing, except to distinguish that car from others. Or a "1" can be used to denote male and a "0," a female. These "values" cannot be ranked, only identified.

2. Ordinal Measurement

This involves not only classifying data into categories, but also being able to order or rank them.[93] The magnitude of the number has a relative meaning in terms of a measurement being larger or smaller than another relative to some explicit value. In our stock car example, when the results of the race are reported (first place, second place, third place), an ordinal scale is established.

3. Interval-Ratio Measurement

In addition to representing a class and a rank, these units of measurement establish a nonsubjective degree of difference between values (how much larger or smaller one measurement is than another).[93] In the stock car race example, suppose the time it took each car to finish the race was recorded. If the fastest car ran a lap in 40 seconds, and the slowest, in 60 seconds, then we could compare the rates of the two cars by saying that the slowest took 20 seconds longer than the fastest.

F. Basic Principles

If an experiment is to be performed in the most efficient way, then a scientific approach to planning the experiment must be employed. The *statistical design of experiments* refers to the process of planning the experiment so that appropriate data will be collected for analysis using specific statistical methods that provide valid and objective conclusions. When the problem involves data that are subject to experimental errors, statistical methodology is the only objective approach to analysis.[78] Thus, there are two aspects to any experimental problem: the design of the experiment and the statistical analysis of the data. These two subjects are obviously closely related, since the method of analysis depends directly on the design employed.

As discussed in detail earlier (Chapter 6), the three basic principles of experimental design are *replication*, *randomization*, and *blocking*. In brief, replication refers to repetition of the basic experiment in order to establish a measurement of, and to correct for experimental error. Randomization requires that both the allocation of the experimental material and the order of the individual runs or trials of the experiment be randomly determined. This is because virtually all statistical methods require that observations (or errors) be independently distributed random variables, and randomization usually makes this assumption valid.[91] Finally, blocking is a technique used to increase the precision of an experiment. A block is one of several groupings ("blocks") of the experimental material that should be more homogeneous than the entire set of material. Blocking involves making comparisons among the conditions of interest in the experiment within each block.

To use the statistical approach in designing and analyzing an experiment, it is necessary that everyone involved in the experiment have a clear idea in advance of exactly what is to be studied, how the data are to be collected, and at least a qualitative understanding of how these data are to be analyzed.[92] A recommended procedural outline follows.

    1. **Recognition of and statement of the problem.** This may seem to be a rather obvious point but, in practice, it is not often simple recognizing that a

problem requiring experimentation exists, nor to develop a clear and generally accepted statement of it. A clear statement of the problem, regardless of how complex or obscure the issues are, is an absolute necessity to devising a solution.

2. **Choice of factors and levels.** The experimenter must select the independent variables or factors to be investigated in the experiment; these may be either quantitative or qualitative. If they are quantitative, thought should be given as to how these factors are to be controlled at the desired values and how they are to be measured. One must also select the ranges over which these factors are to be varied and the number of levels at which evaluations are to be made. These levels may be chosen specifically or selected at random from the set of all possible factor levels.

3. **Selection of a response variable.** In choosing a response or dependent variable, the experimenter must be certain that the response to be measured really provides information about the problem under study. Thought must also be given to how the response will be measured and to the probable accuracy of those measurements.

4. **Choice of experimental design.** This step is of primary importance in the experimental process. The experimenter must determine the degree of difference in true response one wishes to detect and the magnitude of the risks one is willing to tolerate so that an appropriate sample size (number of replicates) may be chosen. At the same time, the order in which the data will be collected and the method of randomization to be employed must be considered. It is always necessary that a good balance between statistical accuracy and its cost be achieved. Most experimental designs used in this book are both statistically efficient and economical. A mathematical model for the experiment must also be devised that matches the statistical analysis to be performed.

In selecting the design, it is important to keep the experimental objectives in mind. In many topical antimicrobial experiments, one already knows at the outset that some of the factors produce different responses. Consequently, one may be interested in identifying *which* factors cause this difference and in estimating the *magnitude* of response change. In other situations, one may be more interested in verifying uniformity. For example, two antimicrobial product configurations A and B may be compared, A being the standard and B, a (hopefully) more effective alternative. The experimenter will then be interested in determining whether there is or is not a difference in antimicrobial effects between the two products.

5. **Performing the experiment.** This is the actual data collection process. The experimenter must carefully monitor the progress of the experiment to ensure that it is proceeding according to plan. Particular attention should be paid to maintenance of randomization, measurement accuracy, and maintaining as uniform an experimental environment as possible.

6. **Data analysis.** In recent years, the computer has played an ever-increasing role in data analysis. There are currently several excellent statistical software packages for analysis of study data. Graphical techniques are particularly helpful in data analysis. An important part of the data analysis process, also available in several packaged forms, is *model-adequacy checking* through the use of exploratory data analysis to examine the underlying statistical model and its associated assumptions.

Remember that statistical methods cannot prove that a factor (or factors) has a particular effect. They only provide guidelines as to the reliability and validity of results. Properly applied, statistical methods do not show anything to be experimentally proven, but they do allow us to measure the likelihood of error in a conclusion or to attach a level of confidence to a statement.

7. **Conclusions and recommendations.** Once the data have been analyzed, the experimenter may draw conclusions or inferences about the results. The statistical inferences must be physically interpreted, and the practical significance of these findings evaluated. Then recommendations concerning these findings must be made. These recommendations may include a further round of experiments, as experimentation is often an iterative process, with one experiment answering some questions and simultaneously posing others. In presenting the results and conclusions, the experimenter should attempt to minimize the use of unnecessary statistical terminology and to phrase information as simply as possible. The use of graphical displays is a very effective way of presenting important experimental results to management.

II. INTRODUCTION TO STATISTICAL INFERENCE

In 1954, there occurred the largest medically-related statistical investigation ever conducted.[80] It dealt with the question: "Does the Salk polio vaccine protect against the polio virus?" To answer this question, more than a million children were randomly assigned to two groups—one to be inoculated with the Salk polio vaccine, the other to be inoculated with a simple salt solution placebo. The data collected from this sample would lead to one of two conclusions: either the vaccine provided immunity to polio or it did not. Such a two-way decisional problem is called a statistical inference. There are two possible outcomes. The test hypothesis (H_A) is true or the null hypothesis (H_O) is true. Both cannot be true and both cannot be false. In this case, the null hypothesis that the vaccine provided no immunity was to be weighed against the test hypothesis that the vaccine was effective.

Two-way decisional problems are common. Examples include practical issues such as the following:

    a. A researcher is interested in testing if a specific presurgical prep product will show greater antimicrobial efficacy than will a control product;

    b. A sociologist suggests that certain environmental changes will reduce the crime rate in a community;

    c. An agronomist predicts greater yield per acre if farmers use a new hybrid seed; and

    d. A new packaging process is claimed to reduce damage to goods shipped by mail order businesses, and so on.

The need to collect valid data either substantiating or disproving the claim of the test hypothesis is a characteristic requisite to all parametric statistics. The evidence one collects on which to base the decision comes from a representational sample set of the entire, larger (often much larger) population. Since sample data must be used, sample variability will be of importance and be referred to as random variability or normal error.

A. Statistical Estimation and Hypothesis

The objective of statistical hypothesis is to validate or disprove an assumption or claim made about a specific parameter of a given population. The evidence one collects to help make the decision comes from the sample data. A statistic (usually the sample mean, $\bar{x}$) computed from one or more samples drawn from a population is used to approximate the value of that parameter for the entire population (usually the population mean, μ). The sampling distribution of $\bar{x}$ (the sample mean) provides an unbiased estimate of μ (the true population mean), if sampling has been random—that is, $\bar{x}$ of the sample set can be considered to equal to μ, or $\bar{x} = \mu$. Likewise, the sample variance (s^2) can be considered an unbiased measurement of the true population variance (σ^2).

In this work, we will simplify this process into a six-step procedure, as follows:

Step 1: Formulate the null hypothesis (H_O) and an alternative hypothesis (H_A); that is, formulate the claim and its negation to serve as the two alternatives being considered ("mutually exclusive").

Step 2: Choose a sample size *n* and a risk factor, the probability of a type I error (α).

Step 3: Choose the test statistic to be used, as appropriate for the sample data.

Step 4: Formulate a decision rule; that is, decide what the evidence must show in order to either support or disprove the test hypothesis.

Step 5: Collect the sample data and perform the chosen statistical calculations.

Step 6: Apply the decision rule and make the decision based upon statistical results.

Thus, a hypothesis is a statement (or claim) about a parameter of a population, and it is our purpose to decide, on the basis of experimental or sample evidence, whether the statement is most probably true or false.

1. Step 1

Step 1 requires the formulation of two hypotheses. The first is the situation currently thought to be true, a statement of the status quo. For example, one may claim (or wish to establish) that:

a. A new prep provides faster antimicrobial effects; or that
b. A new antimicrobial soap for foodhandlers leaves substantially less soap residue on the skin; or that
c. A regular sanitation program reduces food contamination; or that
d. A new precatheter prep procedure reduces hospital-acquired catheter infections; etc.

The researcher often hopes to collect information that substantiates the claim. To maintain scientific integrity, however, s/he must take the position that the new claim is *not* true—that the *status quo persists*. In the four examples above, it is assumed or hypothesized that: (a) the old prep is as good or better; (b) the current soap does not leave any more soap residue on the skin than does the new soap; (c) a regular sanitation regimen does not reduce food contamination; and (d) the prep procedure in use is just as effective as the new one.

We call this current position the *null*, or "no change" position. We demand convincing proof before we reject the null hypothesis. The null hypothesis is denoted by H_O, while the hypothesis which states that a significant difference exists is denoted by H_A or H_1 and is called an alternative (or test) hypothesis. In the present work, we require that both hypotheses be stated in terms of some population parameter. The null and the alternative hypotheses must be mutually exclusive in the sense that both cannot be true and, in most cases, the two hypotheses should be complementary.

In many cases, the null hypothesis and the associated alternative hypothesis take on one of the following formats:

I. Nondirectional, or two-tail test
H_O: (parameter of interest) = (specific value)
H_1: (the given parameter) $\neq$ (given value)
Examples:
1. The mean antimicrobial content is 2% (H_O): $\mu = 0.02$
The mean antimicrobial content is *not* 2% (H_1): $\mu \neq 0.02$

2. The proportion of the healthcare workers who do not perform effective handwashes is 56% (H_O): $\pi = 0.56$

The proportion of the population of healthcare workers who do not perform effective handwashes is not 56% (H_1) $= \pi \neq 0.56$

In this format, the null hypothesis states that the parameter *equals* some number, while the alternative states that it is *not* equal to that value (it is either greater or less than the number).

II. One-directional, or one-tail test

Upper tail focus: H_O (parameter) $\leqq$ (stated value) [claim to be disproved]

H_1 (parameter) > (given value) [claim to be proved]

Lower tail focus: H_O (parameter) $\geqq$ (given value) [claim to be disproved]

H_1 (parameter) < (given value) [claim to be proved]

Examples:

1. A chemist suspects that, due to a stability probe, the mean length of time a surgical scrub formulation is stable is less than 18 months. If this is true, the chemist will add a stronger antioxidant preservative; otherwise, no action will be taken. Here, we are concerned with μ, the mean stability of lots of this surgical scrub product. The chemist claims that $\mu < 18$ months, so H_1: $\mu < 18$ (hypothesis to be proved [what is hoped]), while H_0: $\mu \geqq 18$ (current belief [which the chemist disputes] to be disproved).

2. Product formulations claim that, of two delivery methods for a preoperative prepping product, the mean time for delivery method A (μ_A) is less than the mean time for method B (μ_B). However, delivery method A is more expensive. A large pharmaceutical organization is interested in using delivery method A, but only if it can be demonstrated to be faster than delivery method B. In this problem, we are interested in the statistic "d," where "d" is the difference between means ($d = \mu_A - \mu_B$). The product formulators *hypothesize* that $\mu_A < \mu_B$ or $\mu_A - \mu_B < 0$ or d < 0; hence, the claim to be proved is H_1: $\mu_A < \mu_B$ (or $\mu_A - \mu_B < 0$ or d < 0), while H_0: $\mu_A \geqq \mu_B$ (or $\mu_A - \mu_B \geqq 0$ or d $\geqq 0$) and, if this is true, stay with delivery method B.

3. A researcher has reason to believe that prepping a proposed surgical site for more than 5 minutes causes 52% of the patients to develop a significant rash. If this is true, another product will be used. The parameter of interest in this case is π, the fraction of

all surgical patients who are prepped for more than 5 minutes with the product and develop a rash. The claim to be proved is H_1, that $\pi \geqq 0.52$. The claim to be disproved is H_0, that $\pi < 0.52$.

4. Microbiologists are testing two forms of a product in hopes of increasing their residual antimicrobial properties. The first product is called (L) for lipophilic, and the second product is called (P) for protein-binding. They hypothesize that the mean residual properties for the product L (μ_L) is less than the mean residual properties for the product P (μ_P). If true, product L will be substituted for the product P currently in use. The claim to be proved is H_1, that $\mu_L < \mu_P$, and, therefore, that $\mu_L - \mu_P < 0$. The claim to be disproved is H_0, that $\mu_L > \mu_P$, and, therefore, that $\mu_L - \mu_P \geqq 0$.

5. A biochemist suspects that the mean 99% kill-time for *Staphylococcus aureus* exposed to a standard product is *not* 5 minutes, as it should be in order to use the standard. If it is different than this, the standard must be reformulated. The claim to be disproved is H_0, that $\mu = 5$. The claim to be proved is H_1, that $\mu \neq 5$.

Note that:

1. The equality symbol is always included with H_0.
2. H_1 is a statement that indicates the *opinion of the researcher* as to the actual value of the measured parameter.
3. H_0 is a statement of the commonly accepted belief or standard (which the researcher does not believe, a doubt that probably is the foundation for the entire study).
4. If we wish to show that the stated parameter is *too high* or *too low*, use a *directional test*. If we wish to show the stated value of the parameter is *wrong*, use a *nondirectional test*.

2. Step 2

Type I error occurs when one *rejects* as false a null hypothesis that is actually true. Type II error involves *accepting* as true a null hypothesis that is actually false.[77,91]

A. The object of testing hypotheses is to make correct decisions. When making a decision in the face of uncertainty (from a sample which is, by definition, incomplete information about some larger population), one runs the risk of making an error.

B. The probability of making a Type I error is denoted by α.[92]

$$\alpha = P(\text{type I error}) = P \text{ (rejecting } H_0 \text{ when } H_0 \text{ is true)}$$

The *significance level* of a test is the value of α *selected* beforehand for that test.

Actual conditions	Reject H_0	Accept H_0
H_0 true	Type I error	correct
H_0 false	correct	Type II error

C. All of the problems involving hypothesis-testing have a "bell-shaped" probability distribution governing them. If we consider the total area beneath a curve to be equal to 1.00, then the chosen value of α, together with whether that particular test is one-direction or nondirectional, enables us to divide this area into two parts called, respectively, the *region of rejection* and the *region of non-rejection*.

α = Probability that test statistic is in the rejection region when H_0 is true

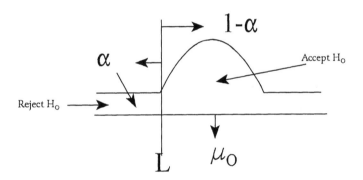

Lower Tail Test.

H_0: $\mu \geq \mu_0$

H_A: $\mu < \mu_0$

A number, L, called the *lower critical value* of α, is determined such that, in repeated sampling from a population with $\mu \geq \mu_0$, the sample mean, $\bar{x}$, would fall below L less than α% of the time. (This same idea applies if the problem involves proportionalities, (e.g., H_0: $\pi \geq \pi_0$ or H_1: $\pi < \pi_0$.)

Upper Tail Test.

H_0: $\mu \leq \mu_0$ or $\pi \leq \pi_0$

H_1: $\mu > \mu_0$ or $\pi > \pi_0$

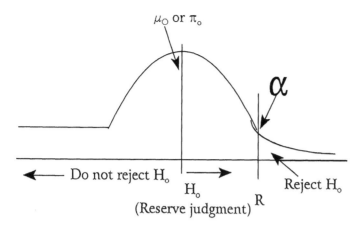

R = *upper critical value*

Two Tail Test.

H_0: $\mu = \mu_0$ or $\pi = \pi_0$

H_1: $\mu \neq \mu_0$ or $\pi \neq \pi_0$

Two numbers, L and R, are found and *one-half* of α is applied to each tail.

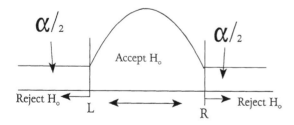

*In each graph, the area represents α.

**L and R are expressed in terms of a number of *standard deviations*.

When a sample calculation falls within the region of rejection, it is said to deviate or differ *significantly* from the stated values of probability of H_0 (the values of L and R found by using the appropriate tables), as we shall see later.

$\beta = P$ *(type II error)* = P *(accepting H_0 when H_0 is false)*
= Probability that sample value falls in region of nonrejection when H_0 is false.

We *choose* a value of α before the test begins. The value of β is *calculated* and depends on: (1) the sample size (n); (2) the chosen value of α; and (3) the specific value of the parameter used *in place of H_0*.

3. Example

Suppose one suspects that a coin is biased in favor of heads. One would like to show that $\pi > \frac{1}{2}$, where π is the proportion of heads if the coin were to be flipped forever. Thus, $H_0 = \pi \leq \frac{1}{2}$ and $H_1 = \pi > \frac{1}{2}$. In this procedure, one requires a definite value of π to be named in the null hypothesis. Because of this, the equality is always included with H_0, and we write $H_0 - \pi = \frac{1}{2}$.

Suppose one decides to toss the coin 20 times and count the number of heads. A look at a standard binomial table where $n = 20$ with $\pi = 0.50$ reveals that the probability of getting 14 or more heads—when $\pi = 0.50$—is less than 5%. When one sets the level of significance, α, at 0.05, one is saying that, *if* one counts 14 or more heads in the sample, one either has a very rare sample *or* there is something wrong with the belief that $\pi = \frac{1}{2}$. The 0.05 is a "cut-off," or critical value. This says that one "expects" 10 heads and 10 tails, but is also aware of sample variation. If 13 or fewer heads result from testing, the difference between what is *expected* and what is *observed* will be attributed to sample variation. If 14 or more heads result, however, one will be alarmed at the discrepancy and state that the difference between what was expected and what was observed is *significant at the $\alpha = 0.05$ level*. One can now formulate a decision rule: (1) toss the coin 20 times; (2) count the number of heads; (3) if one gets 14 or more heads, reject H_0 that $\pi = \frac{1}{2}$; otherwise, do not reject the null hypothesis. By selecting a level of significance of $\alpha = 0.05$, one is saying that one is willing to risk rejecting H_0, even when it is true, 5% of the time. So far, no mention has been made of β—Beta or type II error—which is the probability of accepting H_0 when, in fact, it is false. To calculate β, a *definite* alternative value must be assumed. Suppose one feels that $\pi = 0.70$ for our coin. (The probability of heads is 0.70, so probability of tails is $1 - 0.70 = 0.30$). The probability of 14 or more heads when P(heads) = 0.70 is the same as the probability of 6 or fewer tails when P(tails) = 0.30. Now, consulting a binomial table for $n = 20$ and $\pi = 0.30$, $\beta = P$ (14 or more heads, or $\pi = 0.70$) = P (6 or less tails, or $\pi = 0.30$) = 0.6080. To summarize the decision rule: although the probability of rejecting H_0 when it is true is very low ($\alpha = 5\%$), the probability of failing to reject $\pi = 0.50$ when, in fact, $\pi = 0.70$ is very high ($\beta = 0.61$). The type I error (α) risk is small, but the type II error (β) risk is large. One can *reduce* β by selecting a larger value of α or by taking a larger sample, or both. In general, the type I error is the error one is most concerned with—"the hypothesis $\pi = 0.50$ is assumed true until proven false."

Please note that in this example, a binomial setting was used to facilitate understanding, but the most common distribution setting for the parametric statistics to be discussed will be a Normal (Gaussian) Distribution.

The *test of an hypothesis* is a decision-making procedure consisting of three main parts—formulating the hypothesis to be tested; collecting data relevant to the proving or disproving of the hypothesis; and analyzing the data and making the decision. And the heart of the procedure involves construction of a *confidence interval*, the range of values within which one expects the parameter of interest to be found. The most frequently applied values of α are 0.01, 0.05, and 0.10. If not stated explicitly in a report of results of analysis, $\alpha = 0.05$ in virtually all instances, particularly in the biological sciences.

Problem statements of interest to a researcher take several forms, and are often tested as follows.

A. A test for one mean when standard deviation (σ) is known *or* when $n \geq 30$.

B. A test for one mean when σ is not known and sample is small ($n < 30$).

C. A test for two means when the sample data are independent of each other.

D. A test for two means when samples are paired.

E. A test for one proportion.

F. A test for two proportions.

Using the following common convention for symbols (see table below), let us now apply the six-step procedure to formulate and test hypotheses exemplifying these problem statements.

Population	Measure	Sample of population
N	number	n
μ	mean	$\bar{x}$
σ	standard deviation	s
σ^2	variance	s^2
π	proportion	P

4. Test for One Mean with σ Known

Step 1: Formulate hypothesis (μ is a specific value of μ_o). There are three possibilities relative to the test hypothesis (H_1).

1. H_0: $\mu \leq \mu_o$ versus H_1: $\mu > \mu_o$ (upper tail, directional)*

2. H_0: $\mu \geq \mu_o$ versus H_1: $\mu < \mu_o$ (lower tail, directional)*

3. h_0: $\mu = \mu_o$ versus H_1: $\mu \neq \mu_o$ (two tail, non-directional)*

Step 2: State sample size, n, and select α

Step 3: State the test formula for our test random variable, z.

$z = \dfrac{\bar{x} - \mu_o}{\dfrac{\sigma}{\sqrt{n}}}$ The test formula assumes that the data are *normally*

distributed and if sampling size is with replacement or if N is large. If σ is unknown, but $n \geqq 30$, the formula is altered by replacing σ with s.

Step 4: Formulate a decision rule. State rule in terms of intervals.

Null hypothesis	Test hypothesis	Reject region	Non-reject region
$H_0 = \mu \leqq \mu_0$	$H_1 = \mu > \mu_0$	$z > z_\alpha$	$z \leqq z_\alpha$
$H_0 = \mu \geqq \mu_0$	$H_1 = \mu < \mu_0$	$z < -z_\alpha$	$z \geqq z_\alpha$
$H_0 = \mu = \mu_0$	$H_1 = \mu \neq \mu_0$	$z < -z_{\alpha/2}$ or $z > z_\alpha/_2$	$-z_{\alpha/2} \leqq z \leqq z_{\alpha/2}$

Step 5: Collect sample data and perform statistical calculations.

Step 6: Apply decision rule and make decision

Example 1. The following data represent the number of days a disinfectant solution could be left "mixed" before the antimicrobial efficacy dropped markedly. In 21 independent tests of the product, the following "mixed" shelf-life data are provided: 27, 28, 30, 31, 29, 30, 26, 26, 30, 21, 34, 31, 33, 35, 24, 25, 28, 32, 34, 30, 34. The population standard deviation is $\sigma = 11$ days. If the firm that manufactures the antimicrobial product cannot state that the mean number of days the product is stable when mixed is greater than 34, it will have to reformulate the product. Based on sample information, what should the manufacturer do?

Step 1: $H_1 = \mu > 34$ (what is hoped for by manufacturers [upper tail]; if $\mu = 35$ or 40, or 60 or anything over 34, the product is okay. It is only when $\mu \leqq 34$ that reformulation must occur).

Step 2: For the data above, $n = 21$. Let's pick $\alpha = 0.05$.

Step 3: Because no value for N is given, we assume it is very large and we know $n > 20$.

We will use $z = \dfrac{\bar{x} - \mu_o}{\dfrac{\sigma}{\sqrt{n}}} = \dfrac{\bar{x} - 34}{\dfrac{11}{\sqrt{21}}} = \dfrac{\bar{x} - 34}{2.4}$

Step 4:

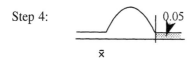

Decision Rule: Find $\bar{x}$, substitute its value into Step 3 and find z; If z > 1.645, reject H_0 and assume product is okay. If $z \leq 1.645$, product is not okay (H_0).

Step 5: Find the mean:

$$\bar{x} = \frac{\Sigma x_i}{n} = \frac{27 + 28 + 30 + \cdots + 30 + 34 + 30}{21} = \frac{618}{21} = 29.43$$

Step 6: Since $-1.90 < 1.645$, do not reject H_0. Conclude that the mean number of days the product is stable is 34, or less. The firm must reformulate.

5. Test for One Mean with σ Unknown

If $n \geq 30$, use the sample procedure just used. For $n < 30$, however, it is better to use the following small sample technique; Steps 1, 2, 5, and 6 are exactly as described in the problem with one mean, σ, known. The formula in Step 3 changes for n < 30 in that s replaces σ. And, in Step 4, each z is replaced by the letter t since z is used to signify a normal distribution, while t indicates Student's t distribution for $n < 30$ (Note: the t and z scales are identical for $n > 30$).

The test formula is distributed as Student's t, with $n - 1$ degrees of freedom (df). Hence, $t = \dfrac{\bar{x} - \mu_o}{s/\sqrt{n}}$, with n − 1 degrees of freedom (df).

Example: Suppose we work the previous example, ignoring the fact that the value of σ was given.

Step 1: $H_0 = \mu \leq 34$ vs. $H_1 = \mu > 34$ (just as in the previous example)

Step 2: $\alpha = 0.05$, $n = 21$ (again, as in the previous example)

Step 3: The test formula $t = \dfrac{\bar{x} - 34}{s/\sqrt{n}}$ with $21 - 1 = 20$ df.

Step 4: A Student's t-table shows that, with 20 df, upper tail area for α = 0.05 = 1.725.

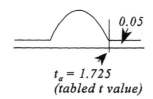

$t_\alpha = 1.725$
(tabled t value)

Decision Rule: (1) take sample; (2) find $\bar{x}$ and s; (3) calculate t; (4) if this $t > 1.725$, reject H_0; otherwise do not reject H_0.

Step 5:
$$\bar{x} = \frac{618}{21} = 29.43$$

$$s^2 = \frac{\Sigma x_i^2 - n(\bar{x})^2}{n-1}$$

$$\Sigma x^2 = 27^2 + 28^2 + 30^2 + 31^2 + 29^2$$
$$+ 30^2 + 26^2 + 26^2 + 30^2 + 21^2$$

$$s^2 = \frac{18460 - 21(29.43)^2}{21 - 1}$$

$$+ 34^2 + 31^2 + 33^2 + 35^2 + 24^2$$
$$+ 25^2 + 28^2 + 32^2 + 34^2$$

$$s^2 = \frac{18460 - 18189}{20}$$

$$+ 30^2 + 34^2 = 18460$$

$$= 13.55$$

$$s = \sqrt{13.55} = 3.68$$

$$\text{thus } t = \frac{\dfrac{29.43 - 34}{3.68}}{\sqrt{21}} = \frac{-4.57}{0.80} = -5.71$$

Step 6:   Since $-5.71 < 1.725$, do not reject H_0.

Example 2.   Consider the following set of $\log_{10}$ measurements: 0.91, 1.32, 1.72, 1.43, 1.66, 1.95, 1.61, 2.07, 2.02, 1.65, 1.52, 1.21, 1.30, 1.59, 1.82, 2.02, 1.69, 2.33, 1.75, 1.46. Could one conclude that the mean of the true population is under 1.70? In order to determine this, apply the five-step process.

Step 1: $H_0 = \mu \geqq 1.70$ vs. $H_1 = \mu < 1.70$ (lower tail).
Step 2: $n = 20$. Let us set $\alpha = 0.025$, for a change.
Step 3:   The test formula is $t = \dfrac{\bar{x} - 1.70}{\dfrac{s}{\sqrt{n}}}$

Step 4:

Decision Rule: (1) take sample; (2) find $\bar{x}$ and s; (3) determine t. If $t < -2.093$, reject H_0; otherwise, do not reject H_0.

Step 5: $\bar{x} = \dfrac{\Sigma x_i}{n} = \dfrac{33.03}{20} = 1.6515,$

$$s = \sqrt{\frac{\Sigma x_i^2 - n(\bar{x})^2}{n-1}} = 0.3346$$

$$t = \frac{1.6515 - 1.70}{\frac{0.3346}{\sqrt{20}}} = 0.6482$$

Since −0.6482 is in the region of non-reject, one cannot reject H_0 at the 0.025 level of significance.

Example 3. A set of 17 observations have $\Sigma x_i = 0.0604$ and $\Sigma x_i^2 = 0.00023124$. Test the hypothesis that $\mu = 0.0038$.

Step 1: $H_0 = \mu = 0.0038$ vs. $H_1 = \mu \neq 0.0038$ (two tail).

Step 2: Given $n = 17$, select $\alpha = 0.01$

Step 3: The test formula is $t = \dfrac{\overline{x} - 0.0038}{\frac{s}{\sqrt{17}}}$ with $17 - 1 = 16$ df.

Use t distribution since σ is unknown and $n < 30$.

Step 4: A Student's t-table shows that with 16df for $\alpha = 0.01$ ($\alpha/2 - 0.005$), the lower tail area $= -2.921$ and the upper tail area $= 2.921$.

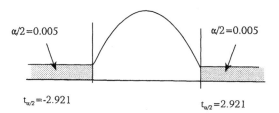

$\alpha/2 = 0.005$ $\alpha/2 = 0.005$

$t_{\alpha/2} = -2.921$ $t_{\alpha/2} = 2.921$

Reject H_0 if $t < -2.921$ or if $t > 2.921$; don't reject H_0 if $-2.921 \leqq t \leqq 2.921$

Step 5: and $\overline{x} = \dfrac{\Sigma x_i}{n} = \dfrac{0.0604}{17} = 0.0035529$

$$s^2 = \frac{\Sigma x_i^2 - n(\overline{x})^2}{n - 1} = \frac{0.00023124 - 17(0.0035529)^2}{16} = 0.00000104$$

Then $s = \sqrt{s^2} = \sqrt{0.00000104} = 0.0010$ and

$$t = \frac{0.0036 - 0.0038}{\frac{0.0010}{\sqrt{17}}} - 0.825$$

Step 6: Since $-2.921 < -0.825 < 2.921$, one cannot reject H_0 at the 0.01 level of significance.

6. Test for Two Means, Independent Samples

Suppose there are two populations, U_1 and U_2, to be compared by an investigator. Let the sample from Population U_1 have size n_1, mean $\bar{x}_1$, and standard deviation s_1. Let the sample from Population U_2 have n_2, mean $\bar{x}_2$, and standard deviation s_2. For now, assume the population variances are equal, $\sigma_1^2 = \sigma_2^2$, although their means may differ. The statistical procedure tests whether it is reasonable to believe that the *difference* between population means is some defined number, k, (often $k = 0$, showing there is no difference). Applying the six-step procedure:

Step 1: Formulate H_0 and H_1. There are three possibilities.
  1) H_0: $\mu_1 - \mu_2 \leq k$ versus H_1: $\mu_1 - \mu_2 > k$ (upper-tail)
  2) H_0: $\mu_1 - \mu_2 \geq k$ versus H_1: $\mu_1 - \mu_2 < k$ (lower-tail)
  3) H_0: $\mu_1 - \mu_2 = k$ versus H_1: $\mu_1 - \mu_2 \neq k$ (two-tail)

Step 2: Choose sample sizes n_1 and n_2, and select value of α.

Step 3: State test formulae:
  A. If n_1 and n_2 are both 30 or more,
  $$z = \frac{(\bar{x}_1 - \bar{x}_2) - k}{\sqrt{\dfrac{s_1^2}{n_1} + \dfrac{s_2^2}{n_2}}}$$

  B. If n_1 and/or n_2 are each less than 30,

  $$t = \frac{(\bar{x}_1 - \bar{x}_2) - k}{\sqrt{s_{pd}^2 \left(\dfrac{1}{n_1} + \dfrac{1}{n_2} \right)}} \text{ with } n_1 + n_2 - 2 \text{ df}$$

  where: $s_{pd}^2 = \dfrac{(n_1 - 1)s_1^2 + (n_2 - 1)s_2^2}{n_1 + n_2 - 2}$

  and s_{pd} = pooled sample variance

Step 4: Formulate a decision rule in terms of intervals.

Step 5: Draw sample and do the calculation.

Step 6: Analyze the data in terms of decision rule and draw conclusion.

Example. Suppose we want to compare the sudsing level of two different antimicrobial soaps. We set up a pilot study using four subjects on the soap formulation currently in use (A) and five other subjects on a new, more expensive formulation (B). The new formulation will be adopted only if there is an average gain of more than five points over the sudsing level of the current formulation.

Step 1: Formulate the null hypothesis, H_0, that the new soap formulation is not at least five points better (stay with the present formulation), and the test hypothesis, H_1, that the new soap formulation is five points greater in sudsing level than the old (put on A).

A) $H_0: \mu_1 - \mu_2 \leq 5$ versus $H_1: \mu_1 - \mu_2 > 5$ (upper-tail)

Step 2: List sample size and α level selection: $n_1 = 4$, $n_2 = 5$, and $\alpha = 0.025$

Step 3: The test formula will use the t distribution, since both n_1 and n_2 are <30. For calculated t (t_c), on the test premise that $\bar{x}_2 > \bar{x}_1 + 5$,

$$t_c = \frac{(\bar{x}_1 - \bar{x}_2) - 5}{\sqrt{s_{pd}^2\left(\frac{1}{n_1} + \frac{1}{n_2}\right)}} \text{ with } n_1 + n_2 - 2 = 4 + 5 - 2 = 7 \text{ df}$$

and where s_{pd}^2 = variance pooled: $S_{p^2d} = \dfrac{(n_1 - 1)s_1^2 + (n_2 - 1)s_2^2}{(n_1 + n_2 - 2)}$

Step 4: A Student's t-table shows that with 7df, the upper tail area for $\alpha = 0.025 = 2.365$.

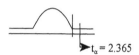

$t_\alpha = 2.365$

Rule: If $t \leq 2.365$, do not reject H_0 and stay with Formulation A
If $t > 2.365$, reject H_0 and use new Formulation B

Step 5: Perform the experiment.

Find $\bar{x}_1$, $\bar{x}_2$, s_1^2, s_2^2

Find t calculated $= t_c = \dfrac{(\bar{x}_1 - \bar{x}_2) - 5}{s_{pd}^2\left(\dfrac{1}{n_1} + \dfrac{1}{n_2}\right)}$

Suppose the following data are gathered:

Soap	Sudsing level				
A (old)	160	173	171	155	
B (new)	181	175	178	185	165

then $\bar{x}_1 = \dfrac{\Sigma x_i}{n_1} = \dfrac{160 + 173 + 171 + 155}{4} = \dfrac{659}{4} = 164.75$

$$\bar{x}_2 = \frac{\Sigma x_i}{n_2} = \frac{181 + 175 + 178 + 185 + 165}{5} = \frac{884}{5} = 176.80$$

$$s_1^2 = \frac{\Sigma x_1^2 - n_i(\bar{x}_1)^2}{(n_1 - 1)} = \frac{[160^2 + 173^2 + 171^2 + 155^2] - 4[164.75]^2}{4 - 1} = 74.9167$$

$$s_2^2 = \frac{\Sigma x_i^2 - n_2(\bar{x}_2)^2}{n_2 - 1} = \frac{[181^2 + 175^2 + 178^2 + 185^2 + 165^2] - 5[176.80]^2}{5 - 1} = 57.2000$$

$$s_{pd}^2 = \frac{(n_1 - 1)s_1^2 + (n_2 - 1)s_2^2}{n_1 + n_2 - 2} = \frac{(4 - 1)(74.9167) + (5 - 1)(57.2000)}{4 + 5 - 2} = 64.7929$$

$$t_c = \frac{(\bar{x}_1 - \bar{x}_2) - 5}{\sqrt{s_{pd}^2\left(\frac{1}{n_1} + \frac{1}{n_2}\right)}} = \frac{(176.80 - 164.75) - 5}{\sqrt{64.7929\left(\frac{1}{4} + \frac{1}{5}\right)}} = \frac{7.05}{\sqrt{64.7929(0.45)}} = \frac{7.05}{5.4} = 1.306$$

$t_c = t$ calculated $= 1.306$

Step 6:   Since 1.306 is less than the critical value of 2.365, do not reject H_0. Stay with Formulation A. Formulation B is not good enough to warrant a switch.

Example.   Two antimicrobial products are compared for antimicrobial efficacy using inhibition zone diameters from multiple agar plates. The results are summarized:

Product 1	Product 2
$n_1 = 36$	$n_2 = 40$
$s_1 = 10.2$	$s_2 = 15.4$
$\bar{x}_1 = 86.03$	$\bar{x}_2 = 92.4$

Are the zones of inhibition of both products statistically equivalent?

Solution.

Step 1:   Determine the type of test (upper tail, lower tail, or two tail). Since we are looking for only a difference, this is a two tail test. H_0: $\mu_1 - \mu_2 = 0$ versus H_1: $\mu_1 - \mu_2 \neq 0$

Step 2:   Thirty-six (36) zones were read for product 1 and forty (40) for product 2, so $n_1 = 36$ and $n_2 = 40$. Pick $\alpha = 0.05$.

Step 3:   Since both n_1 and n_2 are large (greater than 30), we can use the z scale or the normal distribution. $Z = \dfrac{(\bar{x}_1 - \bar{x}_2) - 0}{\sqrt{\dfrac{s_1^2}{n_1} + \dfrac{s_2^2}{n_2}}}$

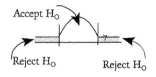

Accept H_O

Reject H_O Reject H_O

Step 4: $z_{\frac{\alpha}{2}} = -1.96$ $z_{\frac{\alpha}{2}} = -1.96$

Decision Rule: If z_c (z calculated) is greater than 1.96 or less than -1.96, reject H_O. If not, accept H_O.

Step 5: Perform the calculations.

$$z = \frac{(86.3 - 92.4) - 0}{\sqrt{\frac{(10.2)^2}{36} + \frac{(15.4)^2}{40}}} = \frac{-6.1}{\sqrt{2.89 + 5.929}} = \frac{-6.1}{2.97} = -2.05$$

Step 6: Since $-2.05 < -1.96$, reject H_O at $\alpha = 0.05$. That is, because -2.05 is well within the region of rejection, we must reject the null (H_O) hypothesis that the two products are equal in antimicrobial efficacy.

7. Test for Difference of Means for Matched or Paired Samples

We wish to test the difference between the means of two populations when the sample observations are related or paired. For example, one could consider a surgical scrub experiment where the test product is evaluated on one hand and a control product on the other, assuming the microbial counts on each hand are equivalent prior to testing. For this study, one could assure the microbial population variances are equal.

Let us look at another matched example. Suppose one wants to know the average zone of inhibition for *Escherichia coli* microorganisms plated on Mueller-Hinton agar. Two product formulations are tested in order to find if they are both as effective. Two zones will be collected from each plate, representing both products.

Plate sample	1	2	3	4	5	6	7	8	9	10
Formulation 1	11.0	5.0	9.8	5.7	6.5	8.2	5.9	6.0	7.5	5.4
Formulation 2	11.1	5.6	9.7	5.3	6.7	8.5	5.6	5.8	7.1	5.5
Difference (d_i)	−0.1	−0.6	0.1	0.4	−0.2	−0.3	0.3	0.2	0.4	−0.1

Do the two methods yield *equal* results at the $\alpha = 0.05$ level?

Step 1: Determine if this is an upper-tail, lower-tail or two-tail test. Since the researcher is interested in only a difference between the two product formulations, this is a two-tail test.

H_o: $\mu_1 - \mu_2 = 0$ (or $D = 0$, where $D = \mu_1 - \mu_2$) versus H_1: $\mu_1 - \mu_2 \neq 0$ (or $D \neq 0$)

Step 2: Select α and determine sample size. Note since the experiment utilized both products on each agar plate, it is a matched pair design. n_1 and $n_2 = 10$, and the α level is set at $\alpha = 0.05$.

Step 3: Since n < 30, use the t distribution, where d = average difference within sample.

$$t = \frac{(\bar{d} - 0)}{\dfrac{S_d}{\sqrt{n}}} \text{ With } 10 - 1 = 9df$$

Step 4: Diagram the acceptance/rejection region for this test.

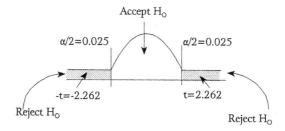

From this diagram, formulate the decision rule.
If $t_c < -2.262$ or $t_c > 2.262$, reject H_0.
If $-2.262 \leq t_c \leq 2.262$, do not reject H_1.

Step 5: Calculate the test statistic from Step 3.

$$\bar{d} = \frac{\Sigma d_i}{n}$$

$$= \frac{(-0.1) + (-0.6) + (0.1) + (0.4) + (-0.2) + (-0.3) + (0.3) + (0.2) + (0.4) + (-0.1)}{10}$$

$$= 0.01$$

$$s_d^2 = \frac{\Sigma d_i^2 - n(\bar{d})^2}{n - 1}$$

$$= \frac{[0.01 + 0.36 + 0.01 + 0.16 + 0.04 + 0.09 + 0.04 + 0.16 + 0.01] - 10(0.01)^2}{10 - 1}$$

$$= 0.1077$$

$$s_d = \sqrt{0.1077} = 0.328$$

$$\frac{\bar{d} - 0}{\dfrac{s_d}{\sqrt{n}}} = \frac{0.01 - 0}{\dfrac{0.328}{\sqrt{10}}} = \frac{0.01}{0.104} = 0.096$$

Step 6: Since $-2.626 < 0.096 < 2.262$, do not reject H_0. The two products do not have significantly different means.

Testing Using Proportions: The above strategy can also be used for proportions or fractional parts of a whole.

Test of a single proportion: Instead of μ being used, π is used by convention to identify the population proportion. The steps in working the statistical design are identical with the previous example.

Step 1: Formulate H_0 and H_1 based on whether the study is upper, lower or a two-tail test. Again, three possibilities exist:
1. H_0: $\pi \leq \pi_0$ versus H_1: $\pi > \pi_0$ (upper tail)
2. H_0: $\pi \geq \pi_0$ versus H_1: $\pi < \pi_0$ (lower tail)
3. H_0: $\pi = \pi_0$ versus H_1: $\pi \neq \pi_0$ (two tail)

Step 2: Select n and α as before.

Step 3: Select the test formula. Let P = sample proportion.

If $n < 30$, then use the t test: $t_c \dfrac{P - \pi_o}{\sqrt{\dfrac{P(1 - P)}{n}}}$ with $n - 1$ *df*

If $n \geq 30$, then use the z test: $z = \dfrac{P - \pi_o}{\sqrt{\dfrac{P(1 - P)}{n}}}$

Step 4: Diagram and formulate the decision in terms of intervals of rejection and non-rejection for the null test hypotheses used in Step 1.

Step 5: Perform the experiment and the statistical calculations.

Step 6: Use decision rule to draw conclusion.

Example. A drug company claims that its new presurgical topical product is effective in at least 80% of the cases where it is used. A sample of 100 patients treated with the new product included 82 who experienced relief. At the 0.05 level of significance, do these data substantiate the firm's claims?

Solution.

Step 1: H_0: $\pi \leq 0.80$ versus H_1: $\pi > 0.80$; this is an upper tail test.

Step 2: $n = 100$, and $\alpha = 0.05$

Step 3: Test formula (since $n \geq 30$)

$$z = \frac{P - 0.80}{\sqrt{\dfrac{P(1 - P)}{n}}}$$

$\underline{P}$ = sample proportion

Step 4:

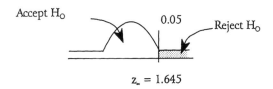

Accept H_O 0.05   Reject H_O

$z_\alpha = 1.645$

Decision Rule:

If $z \leqq 1.645$, do not reject H_0.

If $z > 1.645$, reject H_0 and conclude that the firm's claim is valid.

Step 5: $\underline{P} = \dfrac{82}{100} = 0.82$, $z = \dfrac{0.82 - 0.80}{\dfrac{\sqrt{(0.82)(0.18)}}{100}} = \dfrac{0.02}{0.0384} = 0.521$

Step 6: Since $0.521 < 1.645$, do not reject H_0. The data do not substantiate the firm's claim. 82 of 100 is not too unusual a sample, even if $\pi \leqq 0.80$—it could be a high sample.

8. Testing of Two Proportions

The same six step approach can again be used in comparing two proportions.

Step 1: Formulate the hypothesis:

1) H_0: $\pi_1 - \pi_2 \leqq \pi_0$ versus H_1: $\pi_1 - \pi_2 > \pi_0$ (upper tail)
2) H_0: $\pi_1 - \pi_2 \geqq \pi_0$ versus H_1: $\pi_1 - \pi_2 < \pi_0$ (lower tail)
3) H_0: $\pi_1 - \pi_2 = \pi_0$ versus H_1: $\pi_1 - \pi_2 \neq \pi_0$ (two tail)

Step 2: Determine sample size for each group that, if possible, should be consistent between them. Set α.

Step 3: The test formula is: $z^* = \dfrac{(\underline{P}_1 - \underline{P}_2) - \pi_o}{\sqrt{\dfrac{\underline{P}_1(1 - \underline{P}_1)}{n_1} + \dfrac{\underline{P}_2(1 - \underline{P}_2)}{n_2}}}$

*Note: For sample sizes less than 30, merely substitute the t distribution.

Step 4: Make a graph to establish regions of rejections and non-rejection.

Step 5: Do experiment, collect data, and perform calculations.
Step 6: Use decision rule to draw a statistical conclusion.

Example. The skin-prepping procedures of operating room nurses are under investigation. The claim is made that the proportion of operating room nurses who skip a 10-minute dry-on time is greater in emergency rooms than is the proportion of operating room nurses who do so for elective surgeries. Fifty (50) ER and 75 surgical personnel are observed. Fifteen (15) ER nurses skipped the drying step, and 18 OR nurses did. What do the data signify?

Solution.

Step 1: H_0: $\pi_{ER} - \pi_{OR} \leqq 0$ versus H_1: $\pi_{ER} - \pi_{OR} > 0$ (upper tail)
Step 2: Determine n and set α: $n_{ER} = 50$, $n_{OR} = 75$, and $\alpha = 0.05$
Step 3: Test formula:

$$z = \frac{(P_{OR} - P_{ER}) - 0}{\sqrt{\dfrac{P_{OR}(1 - P_{OR})}{n_1} + \dfrac{P_{ER}(1 - P_{ER})}{n_2}}}$$

Step 4:

$$Z_\alpha = 1.645$$

Decision Rule: If z (calculated) $\leqq 1.645$, do not reject H_0. If $z > 1.645$, reject H_0.

Step 5: Calculate z. $P_{OR}b = \dfrac{15}{50} = 0.03$, $P_{ER} = \dfrac{18}{75} = 0.24$

$$z = \frac{(0.30 - 0.24) - 0}{\sqrt{\dfrac{0.30(0.70)}{50} + \dfrac{(0.24)(0.76)}{75}}} = \frac{0.06}{\sqrt{0.0042 + 0.002432}}$$

$$= \frac{0.06}{\sqrt{0.006632}} = \frac{0.06}{0.081} = 0.74$$

Step 6: Since z calculated is $0.74 < 1.645$, do not reject H_0. The sample evidence does not support the claim that a significant difference exists between ER and OR personnel for post-skin-prep drying times at the 0.05 level of significance.

III. ANALYSIS OF VARIANCE (ANOVA)

To this point, we examined procedures for comparing the mean of one population to the mean of another to decide whether the two populations have essentially the same means or different ones.[77,93,94] We tested these procedures in terms of two special cases: (1) the case in which samples of the two populations are drawn independently, and (2) the case in which paired observations result from the sampling process, including those involving proportionality. However, many situations occur in which one wishes to compare more than two populations.

Suppose one wants to evaluate the antimicrobial efficacies of four different products. One performs a time-kill evaluation using these four products with five replicate runs for each product. The results expressed in $\log_{10}$ reductions from the initial microbial population after 15-second exposures are shown in the table below. The mean reduction for each product is also shown. We note that the rank order by magnitude (of mean log reduction) is Product 2 = 1.4; Product 4 = 1.3; Product 1 = 1.2; and Product 3 = 1.1.

Product 1	Product 2	Product 3	Product 4
1.4	1.5	1.0	1.3
1.1	1.3	1.2	1.1
1.2	1.2	1.1	1.2
1.1	1.6	1.0	1.5
1.1	1.2	1.2	1.3
$\bar{x}_1 = 1.2$	$\bar{x}_2 = 1.4$	$\bar{x}_3 = 1.1$	$\bar{x}_4 = 1.3$

However, the differences among these sample means could be due simply to random error. By noting the data, we see that difference among products may be substantial. In row 1, the data entries go from 1.4 to 1.5 to 1.0 to 1.3, and the same sense of variation is found in the remaining rows. As we look down each column, however, we also note differences in outcomes for each test product. The difference within columns cannot be attributed to actual efficacy differences among the different products. We use these within-product differences as an estimate of our sampling variability or sampling error. We devise a ratio:

$$F = \frac{random\ errors + differences\ between\ products}{random\ error}$$

If there is no difference among the products, this F-ratio would be near 1.[79] This is because the within-product variation of the five replicate samples is equivalent to the between-product differences. When differences among the products are more pronounced, the between-product value becomes larger than the within-product variability and, thus, the overall ratio increases. As the F-ratio increases, we become more convinced that there are significant differences among products. We can establish a critical value, and when the F-ratio exceeds the critical value, we reject our null hypothesis of no significant differences.

A. One-Way Analysis of Variance

For this example, let μ_1, μ_2, μ_3, and μ_4 represent the population means for each product. The null hypothesis is that $\mu_1 = \mu_2 = \mu_3 = \mu_4$, while the test hypothesis is that at least one is different. ANOVA problems are always two tail. We are interested in a difference among groups. The process in solving this problem is called one-way because we consider only one cause of variability: product. Let us designate the following terms: c = number of products being considered, r = number of replicate observations per product, and x_i = individual observations within the data matrix.

$$\bar{x}_j = \frac{\Sigma x_{ij}}{r} = sample\ mean\ for\ product, \ and\ \bar{\bar{x}} = \frac{\Sigma \bar{x}_j}{c} = grand\ mean$$

$$SSA\ (= SST) = r\Sigma(\bar{x}_j - \bar{\bar{x}})^2 = sum\ of\ squares\ between\ products$$

$$SSW\ (= SSE) = \frac{\Sigma\Sigma}{j\ i}[(x_{ij} - \bar{x}_j)^2] = sum\ of\ squares\ within\ each\ product$$

$$(test\ replicates)$$

$$SS = \Sigma(\bar{x}_i - \bar{\bar{x}})^2] = total\ sum\ of\ squares\ within\ each\ factor$$

$$MSA\ (= MST) = \frac{SSA}{c-1}\ (= \frac{SST}{c-1}) = treatment\ mean\ square$$

$$MSE = \frac{SSW}{c(r-1)}\ (= \frac{SSE}{c(r-1)}) = errors\ mean$$

$$F = \frac{variance\ explained\ by\ treatments}{variance\ unexplained} = \frac{MSA}{MSW}\left(\frac{MST}{MSE}\right)$$

In our example:

$$\bar{x}_1 = 1.2,\ \bar{x}_2 = 1.4,\ \bar{x}_3 = 1.1,\ \bar{x}_4 = 1.3\ and\ \bar{\bar{x}} = \frac{1.2 + 1.4 + 1.1 + 1.3}{4} = 1.2$$

$$SSA = r\Sigma(\bar{x}_j - \bar{\bar{x}})^2 = 5[(1.2 - 1.2)^2 + (1.4 - 1.2)^2 + (1.1 - 1.2)^2 + 1.3 - 1.2)^2]$$
$$= 0.30$$

$$SS = \Sigma(\bar{x}_i - \bar{\bar{x}})^2 = (1.4 - 1.2)^2 + (1.1 - 1.2)^2 + (1.2 - 1.2)^2$$
$$+ [all\ columns] + (1.3 - 1.2)^2 = 0.54$$

$$Then\ SSW = SS - SSA = 0.54 - 0.30 = 0.24$$

$$MSA = \frac{SSA}{c-1} = \frac{0.30}{3} = 0.10, \ where \ c = 4, \ and$$

$$MSW = \frac{SSW}{c(r-1)} = \frac{0.24}{16} = 0.015$$

The data can be summarized in an *ANOVA Source Table*:

Source of variation	Degrees of freedom	Sum of squares	Mean of squares	F
Between test products	c − 1 (3)	SSA (0.30)	MSA (0.10)	$\frac{MSA}{MSW} = 6.67$
Within columns	c(r − 1)(16)	SSW (0.24)	MSW (0.015)	
Total	cr − 1(19)	SS(0.54)		

Suppose we pick $\alpha = 0.01$; then, from a standard F-distribution table, we find the critical cut-off $F_{0.01, \, 3, \, 16} = 5.29$.

Since we calculate $F = 6.67$ and $6.67 > 5.29$, we reject H_0 at 0.01 level. The differences are not attributable to sample variation. At least one of the four products differs from the others in efficacy.

Confidence interval for the population mean for each individual product,

$$\mu_j: \ \overline{x}_j - t_{\alpha/2} \sqrt{\frac{MSW}{r}} \leq \mu_j \leq \overline{x}_j + t_{\alpha/2} \sqrt{\frac{MSW}{r}} \ \text{with c(r − 1) df.}$$

B. Two-Way (Factor) ANOVA

In the product example, it is possible that influences other than the specific product formulation had an impact on data.[76] These influences affect the variance within each product, rather than the variance among products. Thus, the error component may have been "inflated" by those outside influences, not allowing us to detect a true difference among products when one really exists. Recall this is a type II or beta (β) error. The original classification is one-way (product efficacy). Suppose the incubation temperature is also considered.

Incubation temperature	Product 1	Product 2	Product 3	Product 4	Total
20	1.4	1.5	1.0	1.3	5.2
25	1.1	1.3	1.2	1.1	4.7
30	1.2	1.2	1.1	1.2	4.7
35	1.1	1.6	1.0	1.5	5.2
37	1.1	1.2	1.2	1.3	4.8
Total	5.9	6.8	5.5	6.4	24.6

To calculate F, find SS and SSA as in the one-way ANOVA.

Then $SSW = SS - SSA - SSR$, where $SSR = c\Sigma(\bar{y}_i - \bar{\bar{x}})^2$, and $\bar{y}_i =$ mean of row i. In this example:

$$\bar{y}_i = \frac{5.2}{4} = 1.30;\ \bar{y}_2 = \frac{4.7}{4} = 1.18;\ \bar{y}_3 = \frac{4.7}{4} = 1.18;$$

$$and\ \bar{y}_4 = \frac{5.2}{4} = 1.30;\ and\ \bar{y}_5 = \frac{4.8}{4} = 1.20$$

$SSR = c\Sigma(\bar{y}_i - \bar{\bar{x}})^2 = 4(1.30 - 1.25)^2 + (1.18 - 1.25)^2 + (1.18 - 1.25)^2 + (1.30 - 1.25)^2 + (1.20 - 1.25)^2 = 0.07$ and $SSW = SS - SSA - SSR + 0.54 - 0.30 - 0.07 = 0.17$

Source of variation	Degrees of freedom	Sum of squares	Mean of squares	F
Between test products	C – 1(3)	SSA (0.30)	MSA (0.10)	$\frac{MSA}{MSW}$ (7.14)
Sum of squares (Rows)	r – 1(4)	SSR (0.07)	MSR (0.018)	$\frac{MSR}{MSW}$ (1.29)
Within columns	(c – 1) (r – 1)(12)	SSW (0.17)	MSW (0.014)	
Total	rc – 1(19)	SS (0.54)		$'$

where $MSR = \dfrac{SSR}{c - 1}$ and $MSW = \dfrac{SSW}{(c - 1)(r - 1)}$

We test two null hypotheses:

1. H_O for columns (products), that $\mu_1 = \mu_2 = \mu_3 = \mu_4$.
2. H_O for rows (temperatures), that $\mu_1 = \mu_2 = \mu_3 = \mu_4 = \mu_5$.

In both cases, the test hypothesis, H_1, is that they are not all equal. From a standard F-distribution table for $\alpha = 0.01$, we find that $F_{0.01,12,3}$ (products) = 5.95 and $F_{0.01,12,4}$ (temperature) = 5.41.

Since F(calculated) for products = 7.14 > 5.95, we reject H_O and conclude that at least one of the products differs from the others in efficacy. However, the F(calculated) for effect of temperature = 1.29 < 5.41, and we cannot reject H_O; incubation temperature caused no significant difference in testing outcomes among the products.

C. Nonparametric Statistics

To this point, we have discussed parametric statistical procedures. A characteristic of parametric procedures is that the appropriateness of their use for purposes of inference depends on certain mathematical assumptions, particularly that sample data are normally distributed and, when comparing two or more groups, that the variances of the sample data are the same.

Since study data do not always meet these assumptions and because research and development funding is usually scarce, requiring that all testing be done on small sample sizes, we frequently need inferential procedures whose validity does not depend on meeting rigid assumptions. Nonparametric statistical procedures fill this need in many instances, since they are valid under the minimal general assumptions that sample data were collected randomly.

D. Advantages of Nonparametric Statistics

The following are some of the advantages of using nonparametric statistical procedures.[83,85,86]

1. Since most nonparametric procedures depend on a minimum of assumptions, the chance of their being improperly used is small.
2. For some nonparametric procedures, the computations can be quickly and easily performed, especially if calculations are done by hand. Thus, using them saves computation time. This can be an important consideration if results are needed in a hurry or if high-powered calculation devices are not available.
3. Researchers with minimum preparation in mathematics and statistics usually find the concepts and methods of nonparametric procedures easy to understand.
4. Nonparametric procedures may be applied when the data are measured on a weak measurement scale, as when only count data or rank data are available for analysis.
5. Sample sizes can be small. However, because nonparametric statistical inferences are generally very conservative, it is difficult to detect a true difference, except when the difference is relatively great. Hence, the probability of type II error is greater than with parametric statistics. Even so, type I error is generally considered the more critical of the two errors (stating there is a difference between groups, when there is not). Nonparametric, as well as parametric, statistics require that the alpha level of significance be determined prior to analysis.

E. Disadvantages of Nonparametric Statistics

Nonparametric procedures are not without disadvantages. The following are some of the more important disadvantages.[80]

1. Because the calculations needed for most nonparametric procedures are simple and rapid, these procedures are sometimes used when parametric procedures are more appropriate. Such a practice often wastes information, since more could be determined from the data using parametric statistics..

2. Although nonparametric procedures have a reputation for requiring only simple calculations, the arithmetic in many instances is tedious and laborious.

F. When to Use Nonparametric Procedures

The following are some situations in which the use of a nonparametric procedure is appropriate.[85,86]

1. The sample sizes are small.
2. The data have been measured on a scale weaker (nominal or ordinal) than that required for the parametric procedure that would otherwise be employed. For example, the data may consist of count data or rank data, thereby precluding the use of some otherwise appropriate parametric procedure.
3. The assumptions necessary for the valid use of a parametric procedure are not met. In many instances, the design of a research project may suggest a certain parametric procedure. Examination of the data, however, reveals that one or more assumptions underlying the test are grossly violated, most often, that the data are non-normally distributed. In such cases, nonparametric procedures are frequently the only alternative.

Let us now discuss a nonparametric analog for the Student's t-test, the Mann-Whitney U Test.

G. Mann-Whitney U Test

We will continue to apply the six step procedure used in our earlier parametric analysis in illustrating use of the Mann-Whitney U test.

Step 1: Formulate the hypothesis:

H_0: Group A is $\leq$ Group B versus H_1: Group A is $>$ Group B (upper tail)

H_0: Group A is $\geq$ Group B versus H_1: Group A is $<$ Group B (lower tail)

H_0: Group A is $=$ Group B versus H_1: Group A is $\neq$ Group B (two tail)

Step 2: Select the α level, as well as the sample sizes: n_A, n_B.

Step 3: Write down the test formula for the Mann-Whitney U test:

$$T = S - \frac{n_1\,(n_1 + 1)}{2}$$

T = test statistic

S = Sum of the ranks assigned to sample Group A

n_1 = sample size of Group A

Step 4: Make decision rule in terms of what T-variables will cause one to accept H_0 and what values will cause one to reject H_0, given the α level selected. Diagram the acceptance/rejection regions, even though a normal distribution representation is not technically correct. The purpose of the diagram is to assist one in keeping a clear picture of what one is doing.

Step 5: Perform experiment and calculate T.

Step 6: Use decision rule from Step 4 to make decision as to accepting or rejecting H_0, based on the calculation from Step 5.

H. Mann-Whitney Test Statistic Discussion

To compute the observed value of the test statistic, we combine the two samples and rank all sample observations from smallest to largest. We assign tied observations the mean of the rank positions they would have occupied had there been no ties. We then sum the ranks of the observations from sample population A (that is, the Xs) and then from sample population B (the Ys). If the location parameter of sample population A is smaller than the location parameter of sample population B, we expect (for equal sample sizes) the sum of the ranks for sample population A to be smaller than the sum of the ranks for sample population B. Similarly, if the location parameter of sample population A is larger than the location parameter of sample population B, we expect just the reverse to be true. The test statistic is based on this rationale in such a way that, depending on the null hypothesis, either a sufficiently small or a sufficiently large sum of ranks assigned to sample observations from sample population A causes us to reject the null (H_0) hypothesis.

1. Decision Rule

The choice of a decision rule depends on the null hypothesis. The possible choices for an α level of significance are as follows.

A. Two tail: When we test H_0 of A, we reject H_0 for either a sufficiently small or a sufficiently large value of T. Therefore, we reject H_0 if the computed value of T is less than $w_{\alpha/2}$ or greater than $w_{1-\alpha/2}$, where

$w_{\alpha/2}$ is the value of T given in Table 1 of Critical Values of t, and $w_{1-\alpha/2}$ is given by $w_{1-\alpha/2} = n_1 n_2 - w_{\alpha/2}$.

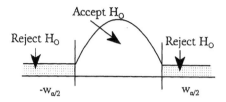

B. Lower tail: When we test H_0 of B, we reject it for sufficiently small values of T. Reject H_0 if the computed T value is less than w_α, the value of T obtained, once again, from Table 1 of Critical Values of t.

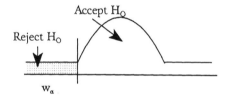

C. Upper tail: When we test H_0 of C, we reject H_0 of C for sufficiently large values of T. Therefore, we reject H_0 if the computed value of T is greater than $w_{1-\alpha}$, where $w_{1-\alpha} = n_1 n_2 - w_\alpha$.

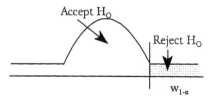

2. Example

A researcher has two antimicrobial products and wants to determine if they differ from one another in zone inhibition diameters on agar plates when challenged with *Escherichia coli* O157, H7. When performing experiments, it is always wise to keep the sample size between the groups the same. In this example, they are not, simply for demonstration purposes.

Product 1	Inhibition zone diameter (X)	Product 2	Inhibition zone diameter (Y)
1	11.9	1	6.6
2	11.7	2	5.8
3	9.5	3	5.4
4	9.4	4	5.1
5	8.7	5	5.0
6	8.2	6	4.3
7	7.7	7	3.9
8	7.4	8	3.3
9	7.4	9	2.4
10	7.1	10	1.7
11	6.9		
12	6.8		
13	6.3		
14	5.0		
15	4.2		
16	4.1		
17	2.2		

We wish to see whether we can conclude on the basis of these data that the two populations represented are significantly different from each other. We let $X_1 = 11.9$, $X_2 = 11.7, \ldots, X_{17} = 2.2$, and $Y_1 = 6.6$, $Y_2 = 5.8, \ldots, Y_{10} = 1.7$. Let $\alpha = 0.05$.

Using the six step procedure, we will set up the statistical test. We are looking for a difference between groups, so it is a two tail test.

Step 1: H_0: Test product group 1 and Test product group 2 are equal in inhibition zone diameters.

H_1: Test product group 1 and Test product group 2. The two populations differ with respect to location.

Step 2: The sample size for Test product group 1 (n_1) is 17. The sample size for Test product group 2 (n_1) is 10. α is 0.05.

Step 3: Establish the test statistic to be used:

$$T = S - \frac{n_1(n_1 + 1)}{2}$$

where S = sum of ranks assigned to the sample scores (sizes of zones of inhibition) for Test Product Group 1.

Step 4: This is a two tail test.

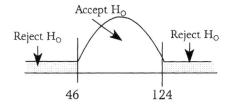

Accept H_O

Reject H_O Reject H_O

46   124

Table 1 shows that $w_{\alpha/2} = 46$ when $n_1 = 17$ and $n_2 = 10$, and $\alpha/2 = 0.05/2$ = 0.025. Thus, by the equation previously discussed for calculating critical values for a two tail test, $w_1 - \alpha/2 = n_1 n_2 - w_{\alpha/2}$, and $w_{1-\alpha/2} = (17)(10) - 46 = 124$.

Decision Rule: Accept H_O if the value of T falls between 46 and 124. Otherwise, reject H_1 at $\alpha = .05$.

Step 5: Perform calculations.

Scores and Corresponding Ranks

Score	Rank	Y Score	Rank
		1.7	1
2.2	2		
		2.4	3
		3.3	4
		3.9	5
4.1	6		
4.2	7		
		4.3	8
5.0	9.5		
		5.0	9.5
		5.1	11
		5.4	12
		5.8	13
6.3	14		
		6.6	15
6.8	16		
6.9	17		
7.1	18		
7.4	19.5		
8.2	19.5		
8.7	21		
8.2	22		
8.7	23		
9.4	24		
9.5	25		
11.7	26		
11.9	27		
Total (= S)	296.5		

Table 1 Quantiles of the Mann-Whitney Test Statistic

	p	$n_2 = 2$ / n_1 = 2	3	4	5	6	7	8	9	10	11	12	13	14	15	16	17	18	19	20
2	.001	0	0	0	0	0	0	0	0	0	0	0	0	0	0	0	0	0	0	0
	.005	0	0	0	0	0	0	0	0	0	0	0	0	0	0	0	0	0	1	1
	.01	0	0	0	0	0	0	0	0	0	0	0	1	1	1	1	1	1	2	2
	.025	0	0	0	0	0	0	1	1	1	1	2	2	2	2	2	3	3	3	3
	.05	0	0	0	0	1	1	2	2	2	2	3	3	4	4	4	4	5	5	5
	.10	0	1	1	2	2	2	3	3	4	4	5	5	5	6	6	7	7	8	8
3	.001	0	0	0	0	0	0	0	0	0	0	0	0	0	0	0	1	1	1	1
	.005	0	0	0	0	0	0	0	1	1	1	2	2	2	3	3	3	3	4	4
	.01	0	0	0	0	0	1	1	2	2	2	3	3	3	4	4	5	5	5	6
	.025	0	0	0	1	2	2	3	3	4	4	5	5	6	7	7	7	8	8	9
	.05	0	1	1	2	3	3	4	5	5	6	6	6	7	9	9	10	10	11	12
	.10	1	2	2	3	4	5	6	6	7	8	9	10	11	11	12	13	14	15	16

	p	$n_2=2$ n_1	3	4	5	6	7	8	9	10	11	12	13	14	15	16	17	18	19	20
4	.001	0	0	0	0	0	0	0	0	1	1	1	2	2	2	3	3	4	4	4
	.005	0	0	0	0	1	1	2	2	3	3	4	4	5	6	6	7	7	8	9
	.01	0	0	0	1	2	2	3	4	4	5	6	6	7	9	8	9	10	10	11
	.025	0	0	1	2	3	4	5	5	6	7	8	9	10	11	12	12	13	14	15
	.05	0	1	2	3	4	5	6	7	8	9	10	11	12	13	15	16	17	18	19
	.10	1	2	4	5	6	7	8	10	11	12	13	14	16	17	18	19	21	22	23
5	.001	0	0	0	0	0	0	1	2	2	3	3	4	4	5	6	6	7	8	8
	.005	0	0	0	1	2	2	3	4	5	6	7	8	8	9	10	11	12	13	14
	.01	0	0	1	2	3	4	5	6	7	8	9	10	11	12	13	14	15	16	17
	.025	0	1	2	3	4	6	7	8	9	10	12	13	14	15	16	18	19	20	21
	.05	1	2	3	5	6	7	9	10	12	13	14	16	17	19	20	22	23	24	26
	.10	2	3	5	6	8	9	11	13	14	16	18	19	21	23	24	26	28	29	31

(Continued)

Table 1 Continued

	p	n_1 = 2	3	4	5	6	7	8	9	10	11	12	13	14	15	16	17	18	19	20
n_2 = 6	.001	0	0	0	0	0	0	2	3	4	5	5	6	7	8	9	10	11	12	13
	.005	0	0	1	2	3	4	5	6	7	8	10	11	12	13	14	16	17	18	19
	.01	0	0	2	3	4	5	7	8	9	10	12	13	14	16	17	19	20	21	23
	.025	0	2	3	4	6	7	9	11	12	14	15	17	18	20	22	23	25	26	28
	.05	1	3	4	6	8	9	11	13	15	17	18	20	22	24	26	27	29	31	33
	.10	2	4	6	8	10	12	14	16	18	20	22	24	26	28	30	32	35	37	39
n_2 = 7	.001	0	0	0	0	1	2	3	4	6	7	8	9	10	11	12	14	15	16	17
	.005	0	0	1	2	4	5	7	8	10	11	13	14	16	17	19	20	22	23	25
	.01	0	1	2	4	5	7	8	10	12	13	15	17	18	20	22	24	25	27	29
	.025	0	2	4	6	7	9	11	13	15	17	19	21	23	25	27	29	31	33	35
	.05	1	3	5	7	9	12	14	16	18	20	22	25	27	29	31	34	36	38	40
	.10	2	5	7	9	12	14	17	19	22	24	27	29	32	34	37	39	42	44	47

$n_2 = 2$	p	n_1 3	4	5	6	7	8	9	10	11	12	13	14	15	16	17	18	19	20
8	.001	0	0	1	2	3	5	6	7	9	10	12	13	15	16	18	19	21	22
	.005	0	0	3	5	7	8	10	12	14	16	18	19	21	23	25	27	29	31
	.01	0	2	5	7	8	10	12	14	16	18	21	23	25	27	29	31	33	35
	.025	1	3	7	9	11	14	16	18	20	23	25	27	30	32	35	37	39	42
	.05	2	5	9	11	14	16	19	21	24	27	29	32	34	37	40	42	45	48
	.10	3	6	11	14	17	20	23	25	28	31	34	37	40	43	46	49	52	55
9	.001	0	0	2	3	4	6	8	9	11	13	15	16	18	20	22	24	26	27
	.005	0	2	4	6	8	10	12	14	17	19	21	23	25	28	30	32	34	37
	.01	0	4	6	8	10	12	15	17	19	22	24	27	29	32	34	37	39	41
	.025	1	5	8	11	13	16	18	21	24	27	29	32	35	38	40	43	46	49
	.05	2	7	10	13	16	19	22	25	28	31	34	37	40	43	46	49	52	55
	.10	3	10	13	16	19	23	26	29	32	36	39	42	46	49	53	56	59	63

(*Continued*)

Table 1 Continued

	p	$n_2=2$ n_1	3	4	5	6	7	8	9	10	11	12	13	14	15	16	17	18	19	20
10	.001	0	0	1	2	4	6	7	9	11	13	15	18	20	22	24	26	28	30	33
	.005	0	1	3	5	7	10	12	14	17	19	22	25	27	30	32	35	38	40	43
	.01	0	2	4	7	9	12	14	17	20	23	25	28	31	34	37	39	42	45	48
	.025	1	4	6	9	12	15	18	21	24	27	30	34	37	40	43	46	49	53	56
	.05	2	5	8	12	15	18	21	25	28	32	35	38	42	45	49	52	56	59	63
	.10	4	7	11	14	18	22	25	29	33	37	40	44	48	52	55	59	63	67	71
11	.001	0	0	1	3	5	7	9	11	13	16	18	21	23	25	28	30	33	35	38
	.005	0	1	3	6	8	11	14	17	19	22	25	28	31	34	37	40	43	46	49
	.01	0	2	5	8	10	13	16	19	23	26	29	32	35	38	42	45	48	51	54
	.025	1	4	7	10	14	17	20	24	27	31	34	38	41	45	48	52	56	59	63
	.05	2	6	9	13	17	20	24	28	32	35	39	43	47	51	55	58	62	66	70
	.10	4	8	12	16	20	24	28	32	37	41	45	49	53	58	62	66	70	74	79

	p	n₂ = 2 n₁	3	4	5	6	7	8	9	10	11	12	13	14	15	16	17	18	19	20
12	.001	0	0	1	3	5	8	10	13	15	18	21	24	26	29	32	35	38	41	43
	.005	0	2	4	7	10	13	16	19	22	25	28	32	35	38	42	45	48	52	55
	.01	0	3	6	9	12	15	18	22	25	29	32	36	39	43	47	50	54	57	61
	.025	2	5	8	12	15	19	23	27	30	34	38	42	46	50	54	58	62	66	70
	.05	3	6	10	14	18	22	27	31	35	39	43	48	52	56	61	65	69	73	78
	.10	5	9	13	18	22	27	31	36	40	45	50	54	59	64	68	73	78	82	87
13	.001	0	0	2	4	6	9	12	15	18	21	24	27	30	33	36	39	43	46	49
	.005	0	2	4	8	11	14	18	21	25	28	32	35	39	43	46	50	54	58	61
	.01	1	3	6	10	13	17	21	24	28	32	36	40	44	48	52	56	60	64	68
	.025	2	5	9	13	17	21	25	29	34	38	42	46	51	55	60	64	68	73	77
	.05	3	7	11	16	20	25	29	34	38	43	48	52	57	62	66	71	76	81	85
	.10	5	10	14	19	24	29	34	39	44	49	54	59	64	69	75	80	85	90	95

(Continued)

Table 1 Continued

n_2	p	n_1=2	3	4	5	6	7	8	9	10	11	12	13	14	15	16	17	18	19	20
14	.001	0	0	2	4	7	10	13	16	20	23	26	30	33	37	40	44	47	51	55
	.005	0	2	5	8	12	16	19	23	27	31	35	39	43	47	51	55	59	64	68
	.01	1	3	7	11	14	18	23	27	31	35	39	44	48	52	57	61	66	70	74
	.025	2	6	10	14	18	23	27	32	37	41	46	51	56	60	65	70	75	79	84
	.05	4	8	12	17	22	27	32	37	42	47	52	57	62	67	72	78	83	88	93
	.10	5	11	16	21	26	32	37	42	48	53	59	64	70	75	81	86	92	98	103
15	.001	0	0	2	5	8	11	15	18	22	25	29	33	37	41	44	48	52	56	60
	.005	0	3	6	9	13	17	21	25	30	34	38	43	47	52	56	61	65	70	74
	.01	1	4	8	12	16	20	25	29	34	38	43	48	52	57	62	67	71	76	81
	.025	2	6	11	15	20	25	30	35	40	45	50	55	60	65	71	76	81	86	91
	.05	4	8	13	19	24	29	34	40	45	51	56	62	67	73	78	84	89	95	101
	.10	6	11	17	23	28	34	40	46	52	58	64	69	75	81	87	93	99	105	111

n_2	p	$n_2=2$ n_1	3	4	5	6	7	8	9	10	11	12	13	14	15	16	17	18	19	20
16	.001	0	0	3	6	9	12	16	20	24	28	32	36	40	44	49	53	57	61	66
	.005	0	3	6	10	14	19	23	28	32	37	42	46	51	56	61	66	71	75	80
	.01	1	4	8	13	17	22	27	32	37	42	47	52	57	62	67	72	77	83	88
	.025	2	7	12	16	22	27	32	38	43	48	54	60	65	71	76	82	87	93	99
	.05	4	9	15	20	26	31	37	43	49	55	61	66	72	78	84	90	96	102	108
	.10	6	12	18	24	30	37	43	49	55	62	68	75	81	87	94	100	107	113	120
17	.001	0	1	3	6	10	14	18	22	26	30	35	39	44	48	53	58	62	67	71
	.002	0	3	7	11	16	20	25	30	35	40	45	50	55	61	66	71	76	82	87
	.01	1	5	9	14	19	24	29	34	39	45	50	56	61	67	72	78	83	89	94
	.025	3	7	12	18	23	29	35	40	46	52	58	64	70	76	82	88	94	100	106
	.05	4	10	16	21	27	34	40	46	52	58	65	71	78	84	90	97	103	110	116
	.10	7	13	19	26	32	39	46	53	59	66	73	80	86	93	100	107	114	121	128

(Continued)

Table 1 Continued

	p	$n_2 = 2$ / n_1	3	4	5	6	7	8	9	10	11	12	13	14	15	16	17	18	19	20
18	.001	0	1	4	7	11	15	19	24	28	33	38	43	47	52	57	62	67	72	77
	.005	0	3	7	12	17	22	27	32	38	43	48	54	59	65	71	76	82	88	93
	.01	01	5	10	15	20	25	31	37	42	48	54	60	66	71	77	83	89	95	101
	.025	3	8	13	19	25	31	37	43	49	56	62	68	75	81	87	94	100	107	113
	.05	5	10	17	23	29	36	42	49	56	62	69	76	83	89	96	103	110	117	124
	.10	7	14	21	28	35	42	49	56	63	70	78	85	92	99	107	114	121	129	136
19	.001	0	1	4	8	12	16	21	26	30	35	41	46	51	56	61	67	72	78	83
	.005	1	4	8	13	18	23	29	34	40	46	52	58	64	70	75	82	88	94	100
	.01	2	5	10	16	21	27	33	39	45	51	57	64	70	76	83	89	95	102	108
	.025	3	8	14	20	26	33	39	46	53	59	66	73	79	86	93	100	107	114	120
	.05	5	11	18	24	31	38	45	52	59	66	73	81	88	95	102	110	117	124	131
	.10	8	15	22	29	37	44	52	59	67	74	82	90	98	105	113	121	129	136	144

	p	$n_2 = 2$ n_1	3	4	5	6	7	8	9	10	11	12	13	14	15	16	17	18	19	20
20	.001	0	1	4	8	13	17	22	27	33	38	43	49	55	60	66	71	77	83	89
	.005	1	4	9	14	19	25	31	37	43	49	55	61	68	74	80	87	93	100	106
	.01	2	6	11	17	23	29	35	41	48	54	61	68	74	81	88	94	101	108	115
	.025	3	9	15	21	28	35	42	49	56	63	70	77	84	91	99	106	113	120	128
	.05	5	12	19	26	33	40	48	55	63	70	78	85	93	101	108	116	124	131	139
	.10	8	16	23	31	39	47	55	63	71	79	87	95	103	111	120	128	136	144	152

$$T = S - \frac{n_1(n_1 + 1)}{2} = 296.5 - \frac{17(17 + 1)}{2} = 143.5$$

Step 6:   Since T = 143.5 is greater than the cut-off value of 124, we reject the H_0 hypothesis at $\alpha = 0.05$. The products differ significantly in their effects.

Large-Sample Approximation.   When either n_1 or n_2 is greater than 20, we cannot use Table 1 of Critical Values of *t*. When n_1 and n_2 are both large, however, the normal distribution applies.[85] Then:

$$z = \frac{T - n_1 n_2/2}{\sqrt{n_1 n_2(n_1 + n_2 + 1)/12}}$$

has approximately the standard normal distribution when H_o is true.

where $T = S - \frac{n_1(n_1 + 1)}{2}$; S = Sum of assigned ranks for Sample 1; n_1 = sample size of Sample 1 and n_2 = sample size of Sample 2. The expected value of T, then, is $n_1 n_2/2$, and its variance is $n_1 n_2(n_1 + n_2 + 1)/12$.

Ties may occur within groups or between them. Ties within groups have no effect on the test statistic, but those between groups do.[83] When we use the large-sample approximation, we may adjust the formula, for the test statistic. Let *t* be the number of ties between groups for a given rank. Then the correction for ties is:

$$\frac{n_1 n_2(\Sigma t^3 - \Sigma t)}{12(n_1 + n_2[n_1 + n_2 - 1])}$$

Hence, the final equation for *z* becomes:

$$\frac{T - n_1 n_2/2}{\sqrt{\frac{n_1 n_2(n_1 + n_2 + 1)}{12} - \frac{n_1 n_2(\Sigma t^3 - \Sigma t)}{12(n_1 + n_2)(n_1 + n_2 + 1)}}}$$

IV.   KRUSKAL-WALLIS STATISTIC

Perhaps the most widely used nonparametric analog of the one-factor analysis of variance for comparing more than two groups is the Kruskal-Wallis one-way analysis of variance by ranks. The Kruskal-Wallis test uses more information than does the Median test, another potential approach to the data.[85,86] As a consequence, the Kruskal-Wallis test is usually more powerful and is preferred when the available data are measured on at least the ordinal scale. When only two samples are being considered, the Kruskal-Wallis test is equivalent to the Mann-Whitney U test discussed above.

Again we will use the six step procedure.

Step 1: Formulate the hypothesis (which will always be two-tail).

H_0: Group A = Group B = Group C = . . . Group K

H_1: At least one group differs from the others.

Step 2: Select α level and sample sizes (n_k) of the test groups. As before, the sample sizes preferably should be equal, but the test can be computed and used even when they are not.

Step 3: Establish the test statistic to be used.

$$H = \frac{12}{N(N+1)} \sum_{i=1}^{k} \frac{R_i^2}{n_i} - 3(N+1)$$

Where: H = Kruskal-Wallis test statistic

N = Total of the sample sizes for the K samples

$= n_1 + n_2 + n_3 + \ldots + n_k$

R_i = Sum of the ranks of the values for K samples

Step 4: Make decision rule in terms of the H-value, and what values of H cause rejection of H_0 and acceptance of H_1

Step 5: Conduct the experiment, collect data and compute the H-value.

Step 6: Use the decision rule from Step 4 as to the acceptance or rejection of H_0 based upon the calculated H-value.

A. Discussion

Data available for analysis may be displayed in a table such as that presented below. We replace each original observation by its rank relative to all the observations in the k samples. If we let $N = \sum_{i=1}^{k} n_i$ be the total number of observations in the k samples, we assign the rank 1 to the smallest of these, rank 2 to the next in size, and so on to the largest, which is given the rank, N. In the case of a tie, we assign the tied observations the average of the ranks that would be assigned if there were no ties.

Data display for Kruskal-Wallis one-way analysis of variance by ranks

Sample			
1	2	. . .	k
$X_{1,1}$	$X_{2,1}$	. . .	$X_{k,1}$
$X_{1,1}$	$X_{2,1}$	. . .	$X_{k,2}$
.	.	.	.
.	.	.	.
.	.	.	.
$X_{1,n1}$	$X_{2,n2}$	. . .	$X_{k,nk}$

If the null hypothesis is true, we expect the distribution of ranks over the groups to be a matter of chance, so that either the small ranks or the large ranks do not concentrate in one sample. Therefore, if the null hypothesis is true, we expect the k sums of ranks (that is, the sums of the ranks in each sample) to be about equal when adjusted for unequal sample sizes. An intuitively appealing test statistic is one that determines whether the sums of the ranks are sufficiently disparate that they are not likely to have been derived from samples from identical populations—leading to rejection of H_o—or whether they are so close in magnitude that we cannot discredit the hypothesis of identical population distributions. Just such a statistic is the Kruskal-Wallis test statistic. It is a weighted sum-of-squares of deviations of sums-of-ranks from the expected sum-of-ranks, using reciprocals of sample sizes as the weights. The formal Kruskal-Wallis test statistic is:

$$H = \frac{12}{N(N+1)} \sum_{i=1}^{k} \frac{1}{N_1} [R_i - \frac{n_i(N+1)}{2}]^2$$

where R_i is the sum of the ranks assigned to observations in the ith sample, and $n_i(N+1)/2$ is the expected sum-of-rank. A computationally more convenient form of this Equation is:

$$H = \frac{12}{N(N+1)} \sum \frac{R_i^2}{n_i} - 3(N+1)$$

B. Decision Rule

When we are considering three samples, and each sample has 5 or fewer observations, we compare the computed value of H for significance with the tabulated value of the test statistic given in Table 2. When the number of samples and/or observations per sample are such that we cannot use Table 2 ($n_i > 5$), we compare the computed value of H for significance with a table of the Chi-square values for $k - 1$ degrees of freedom. We do so since Kruskal shows that, for large n_i (n per group over 30) and k, H is distributed approximately as Chi-square with $k - 1$ degrees of freedom.

1. Example

A researcher is interested in the antimicrobial effects of a topical antimicrobial product in three different formulations, Configurations A, B, and C. A screening evaluation was performed, measuring the diameters of zones of inhibition. The collected data are presented below.

Diameter of inhibition zones in millimeters									
Group A	2.6	3.1	2.1	3.2	4.5	3.3	3.0	1.5	2.9	3.6
Group B	4.6	5.0	4.5	3.5	4.7	3.7	—	—	—	—
Group C	3.4	7.7	2.0	10.5	8.4	6.9	—	—	—	—

The researcher wishes to know whether these data provide sufficient evidence to indicate a difference exists between the three groups in median zone diameters.

Step 1: Establish the hypothesis.

H_o: The three antimicrobial formulations are identical in inhibition zone diameter.

H_1: At least one of the three antimicrobial products is different from the others.

Step 2: In this study, we will set α at the 0.01 level of significance. The sample size for the three product configuration groups are: $n_A = 10$, $n_B = 6$, $n_C = 6$.

Step 3: The test statistic to be used is:

$$H = \frac{12}{N(N+1)} \sum_{i=1}^{k} \frac{R_i^2}{n_i} - 3(N+1)$$

Step 4: This is a two tail test. Since the sample sizes all exceed 5, we cannot use Table 2 but must use a table of Chi-square values to determine whether the sample medians are significantly different. The critical value of Chi-square (χ^2) for $k - 1$ ($= 3 - 1$) $= 2$ degrees of freedom is 9.210 for $\alpha = 0.01$, where k is the number of test groups.

Table 2

Sample sizes					Sample sizes				
n_1	n_2	n_3	Critical value	α	n_1	n_2	n_3	Critical value	α
2	1	1	2.7000	0.500				4.7000	0.101
2	2	1	3.6000	0.200	4	4	1	6.6667	0.010
2	2	2	4.5714	0.067				6.1667	0.022
			3.7143	0.200				4.9667	0.048
3	1	1	3.2000	0.300				4.8667	0.054
3	2	1	4.2857	0.100				4.1667	0.082
			3.8571	0.133				4.0667	0.102
3	2	2	5.3572	0.029	4	4	2	7.0364	0.006
			4.7143	0.048				6.8727	0.011
			4.5000	0.067				5.4545	0.046
			4.4643	0.105				5.2364	0.052
3	3	1	5.1429	0.043				4.5545	0.098
			4.5714	0.100				4.4455	0.103
			4.0000	0.129	4	4	3	7.1439	0.010
3	3	2	6.2500	0.011				7.1364	0.011
			5.3611	0.032				5.5985	0.049
			5.1389	0.061				5.5758	0.051
			4.5556	0.100				4.5455	0.099
			4.2500	0.121				4.4773	0.102

Sample sizes					Sample sizes				
n_1	n_2	n_3	Critical value	α	n_1	n_2	n_3	Critical value	α
3	3	3	7.2000	0.004	4	4	4	7.6538	0.008
			6.4889	0.011				7.5385	0.011
			5.6889	0.029				5.6923	0.049
			5.6000	0.050				5.6538	0.054
			5.0667	0.086				4.6539	0.097
			4.6222	0.100				4.5001	0.104
4	1	1	3.5714	0.200	5	1	1	3.8571	0.143
			4.5333	0.097				4.9091	0.053
			4.4121	0.109				4.1091	0.086
5	4	1	6.9545	0.008				4.0364	0.105
			6.8400	0.011	5	5	2	7.3385	0.010
			4.9855	0.044				7.2692	0.010
			4.8600	0.056				5.3385	0.047
			3.9873	0.098				5.2462	0.051
			3.9600	0.102				4.6231	0.097
5	4	2	7.2045	0.009				4.5077	0.100
			7.1182	0.010	5	5	3	7.5780	0.010
			5.2727	0.049				7.5429	0.010
			5.2682	0.050				5.7055	0.046
			4.5409	0.098				5.6264	0.051

Table 2 Continued

Sample sizes					Sample sizes				
n_1	n_2	n_3	Critical value	α	n_1	n_2	n_3	Critical value	α
			4.5182	0.101				4.5451	0.100
5	4	3	7.4449	0.010				4.5363	0.102
			7.3949	0.011	5	5	4	7.8229	0.010
			5.6564	0.049				7.7914	0.010
			5.6308	0.050				5.6657	0.049
			4.5487	0.099				5.6429	0.050
			4.5231	0.103				4.5229	0.099
5	4	4	7.7604	0.009				4.5200	0.101
			7.7440	0.011	5	5	5	8.0000	0.009
			5.6571	0.049				7.9800	0.010
			5.6176	0.050				5.7800	0.049
			4.6187	0.100				5.6600	0.051
			4.5527	0.102				4.5600	0.100
5	5	1	7.3091	0.009				4.5000	0.102

In two-tail tests for ANOVA, including nonparametric statistics, we do not get lower and higher cut-off values. Instead, we get one value which, in this case, is 9.210. Hence, if the H-value is greater than the χ^2 value of 9.210, we reject the H_O and conclude that a difference exists between the three groups at the $\alpha = 0.01$ level of significance.

Step 5: We calculate the H-value, the test statistic. The ranks replacing the original observations and the sums of ranks are displayed below.

Ranks Corresponding to Zone Size Diameter

Product configuration	Ranks										Rank
Group A	4	7	3	8	14	9	6	1	5	12	$R_A = 69$
Group B	16	18	15	11	17	13	—	—	—	—	$R_B = 90$
Group C	10	20	2	22	21	19	—	—	—	—	$R_C = 94$

From these data, compute H.

$$H = \frac{12}{22(22+1)} [\frac{69^2}{10} + \frac{90^2}{6} + \frac{94^2}{6}] - 3(22+1) = 9.232$$

Step 6:   Decision. Since H = 9.232 is larger than χ^2 (9.210), we reject H_0 at the $\alpha = 0.01$ level of significance. We conclude that the median zone of inhibition of at least one product configuration is different from the others.

Correction for ties: If there are more than three ties in rank, the test statistic should be adjusted.[83,86] The adjustment factor is:

$$1 - \frac{\Sigma T}{N^3 - N}$$

where: $T = t^3 - t$, and t is the number of tied observations in a tied group of scores; N is the number of observations in all k samples together.

The adjusted test statistic becomes:

$$H(adjusted) = H_A = \frac{H}{1 - \Sigma T/(N^3 - N)}$$

The effect of the adjustment is to inflate the value of the test statistic. Thus, if H is significant at the desired level of α without the adjustment, there is no point in computing H_A. Furthermore, Kruskal and Wallis point out that with 10 or fewer samples, a value for H of 0.01 or more does not change more than 10% when the adjusted value of H_A is computed, provided that no more than one-fourth of the observations are involved in ties.[85]

C.   Multiple Comparisons

When a difference is detected between test groups, the researcher will more than likely want to find out where the difference is. The formula to use when the k groups have the same sample size is:

$$|\overline{R}_i - \overline{R}_j| \le z \frac{\sqrt{k(N+1)}}{6}$$

where: $\overline{R}$ = the average rank value per group, $\frac{\Sigma R}{n}$, for any two groups, i and j.

z = critical z value from table below
k = number of groups tested
$n = \Sigma n_i$

Values for Multiple Comparisons of k-Sample

			Level of Significance			
k	0.30	0.25	0.20	0.15	0.10	0.05
2	1.036	1.150	1.282	1.440	1.645	1.960
3	1.645	1.732	1.834	1.960	2.128	2.394
4	1960	2.037	2.128	2.241	2.394	2.638
5	2.170	2.241	2.326	2.432	2.576	2.807

When the sample sizes between the k groups are not equal, the test formula becomes:

$$|\overline{R}_i - \overline{R}_j| \le z \sqrt{\frac{N(N+1)}{12} \left(\frac{1}{n_i} + \frac{1}{n_j} \right)}$$

1.   Example

Let us compare configuration groups to find where the difference exists:

$$|\overline{R}_i - \overline{R}_j| \le z \sqrt{\frac{N(N+1)}{12} \left(\frac{1}{n_i} + \frac{1}{n_j} \right)}$$

Group A vs. Group B

$$\overline{R}_A = \frac{4+7+3+8+14+9+6+1+5+12}{10} = \frac{69}{10} = 6.9$$

$$\overline{R}_B = \frac{90}{6} = 15$$

Let $\alpha = 0.05$ in this case and $k = 3$ (3 groups); therefore,
$z = 2.394$ from the Table of Critical Values, and

$$|6.9 - 15| \leq 2.394 \sqrt{\frac{22(22+1)}{12} \left(\frac{1}{10} + \frac{1}{6}\right)}$$

$8.10 \nleq 8.03$, so Group A and Group B differ at the 0.05 level of significance.

The other group comparisons are calculated in the same manner. Note that there will always be $\frac{k(k-1)}{2}$ possible group combinations to test.

References

1. D. S. Paulson, Quality Assurance of Topical Antimicrobials, Pharmaceutical & Cosmetic Quality, Nov/Dec, 26–32, 1997.
2. D. S. Paulson, Research Designs for the Soaps and Cosmetic Industry: A Basic Approach, Soap/Cosmetics/Chemical Specialties, Nov, 50–58, 1995.
3. K. Wilber. Sex, Ecology, Spirituality. Shambhala, Boston, 1995.
4. K. Wilber. A Brief History of Everything. Shambhala, Boston, 1996.
5. K. Wilber. The Eye of Spirit. Shambhala, Boston, 1997.
6. D. M. Newman. Sociology. Sage, Thousand Oaks, CA, 1997.
7. H. R. Moskowitz. Cosmetic Product Testing: A Modern Psychospiritual Approach, Vol. III, Marcel Dekker, Inc., New York, 1984.
8. R. Searle. The Construction of Social Reality. Free Press, New York, 1995.
9. D. S. Paulson. Successfully Marketing Topical Antimicrobial Products, Soaps/Cosmetics/Chemical Specialities, Feb, 32–38, 1991.
10. D. O. Sears, L. A. Peplau, S. E. Taylor. Social Psychology, 7th ed. McGraw-Hill, Englewood Cliffs, NJ, 1991.
11. R. Kegan. In Over Our Heads. Harvard, Cambridge, 1994.
12. W. C. Frazier & D. C. Westhoff. Food Microbiology, 4th ed.. McGraw-Hill, New York, 1988.
13. D. S. Paulson. Foodborne Disease: Controlling the Problem, Environmental Health. May, 15–19, 1997.
14. D. S. Paulson. Designing a Handwash Efficacy Program, Pharmaceutical & Cosmetic Quality, Jan/Feb, 42–44, 1997.
15. D. S. Paulson. Efficacy Evaluation of a 4% Chlorhexidine Gluconate Solution as a Full-Body Shower Wash, American Journal of Infection Control, 21:4, 205–209, 1993.
16. T. Peters. Liberation Management. Knopf, New York, 1992.
17. C. B. Pert. Molecules of Emotion. Scribner, New York, 11, 1997.
18. J. Neter & W. Wasserman. Applied Linear Statistical Models. Irwin, Homewood, IL, 1974.
19. D. S. Paulson. Comparative Evaluation of Five Surgical Scrub Formulations, Association of Operating Room Nurses Journal, 60:2, 246–256, 1994.
20. D. S. Paulson. Designing a Handwash for Healthcare Workers, Soap/Cosmetics/Chemical Specialties, June, 1996.

21. D. S. Paulson. To Glove or To Wash: A Current Controversy, Food Quality, June/ July, 60–63, 1996.

22. D. S. Paulson. A Proposed Evaluation Method for Antimicrobial Hand Soaps, Soap/ Cosmetics/Chemical Specialties, June, 64–67, 1996.

23. M. J. Marples. The Ecology of Human Skin. Charles Thomas, Springfield, IL, 1965.

24. C. R. Leeson & T. S. Leeson. Histology, 3rd ed. W. B. Saunders, Philadelphia, 1976.

25. W. Montagna, A. M. Kligman, & K. S. Carlisle. Atlas of Normal Human Skin, Springer-Verlag, New York, 1992.

26. F. N. Maarzulli & H. I. Maibach. Dermatoxicology, 3rd ed. Hemisphere Publishing, New York, 1987.

27. H. I. Maibach & R. Aly. Skin Microbiology. Springer-Verlag, New York, 1981.

28. H. Schaefer & T. E. Redelmeier. Skin Barrier. Karger, Basel, Switzerland, 1996.

29. H. Mukhtar. Pharmacology of the Skin. CRC Press, Boca Raton, FL, 1992.

30. M. Schaechter, G. Medoff, & B. I. Eisenstein. Mechanisms of microbial disease, 2nd ed. Williams & Williams, Baltimore, 1993.

31. C. A. Mims. The Pathogenesis of Infectious Disease, 3rd ed. Academic Press, New York, 1987.

32. D. S. Paulson. A Broad-Based Approach to Evaluting Topical Antimicrobial Products, Handbook of Disinfectants and Antiseptics, J. M. Ascenzi, Ed., Marcel Dekker, Inc., New York, 1996.

33. D. S. Paulson. Developing Effective Topical Antimicrobials, Soaps/Cosmetics/ Chemical Specialties, Dec, 50–58, 1997.

34. D. G. Maki. Infections Caused by Intravascular Devices Used for Infusion Therapy: Pathogenesis, Preventions, and Management, In: Infections Associated with Indwelling Medical Devices, A. L. Bisno & F. A. Waldvogal, Eds., American Society for Microbiology Press, Washington, D.C., 155–212, 1994.

35. W. K. Joklik, H. P. Willet, D. B. Amos, & C. M. Wilfert. Zinsser Microbiology, 20th ed. Appleton & Lange, Norwalk, CN, 387–400, 1992.

36. D. H. Groeschel & T. L. Pruett. Surgical Antiseptics. In: Disinfection, Sterilization, and Preservation, 4th ed. S. S. Block, ed., Lea & Febiger, Philadelphia, 1991.

37. D. S. Paulson. Designing a Healthcare Personnel Handwash, Soap/ Cosmetics/ Chemical Specialties, August, 53–57, 1988.

38. L. E. Hood, I. L. Weissman, W. B. Wood, & J. H. Wilson. Immunology, 2nd ed. Benjamin/Cummings, Menlo Park, 1984.

39. W. L. Weissman, L. E. Hood, & W. B. Wood. Essential Concepts in Immunology, Benjamin/Cummings, Menlo Park, 1978.

40. J. Klein. Immunology: The Science of Self-Nonself Discrimination, John Wiley, New York, 1982.

41. J. Kuba. Immunology, 2nd ed. W. H. Freeman, San Francisco, 1991.

42. M. S. Thaler, R. D. Klausner, & H. J. Cohen. Medical Immunology, J. B. Lippincott, Philadelphia, 1977.

43. B. Benacerraf & E. R. Unanue. Textbook of Immunology, Williams & Williams, Philadelphia, 1979.

44. E. J. Baron, L. R. Peterson, & S. M. Feinegold. Bailey and Scott's Diagnostic Microbiology, 9th ed. Mosby, St. Louis, 1994.

45. T. D. Brock, D. W. Smith, & M. T. Madigan. Biology of Microorganisms, 4th ed. Prentice-Hall, Englewood Cliffs, NJ, 1984.

46. R. M. Atlas. Handbook of Microbiological Media, CRC Press, Boca Raton, 1993.

47. B. A. Freeman. Burrows Textbook of Microbiology, 22nd ed. W. E. Saunders, Philadelphia, 1985.

48. R. Y. Stanier, E. A. Adelberg, & J. Ingraham. The Microbial World, 4th ed. Prentice-Hall, Englewood Cliffs, NJ, 1976.

49. M. J. Pelezar, R. D. Reid, & E. C. S. Chan. Microbiology, McGraw-Hill, 1977.

50. P. R. Murray, E. J. Baron, M. A. P. Fallen, F. C. Tenover, & R. H. Yolken. Manual of Clinical Microbiology, 6th ed. ASM Press, Washington, D. C., 1995.

51. P. L. Carpenter. Microbiology, 4th ed. W. B. Saunders, Philadelphia, 1977.

52. B. D. Davis, R. Dulbecco, H. N. Eisen, & H. S. Ginsberg. Microbiology, 3rd ed. Harper & Row, Cambridge, 1980.

53. J. D. Watson, N. H. Hopkins, J. W. Roberts, J. A. Steitz, & A. M. Weiner. Molecular Biology of the Gene, Vol 1, 4th ed. Benjamin/Cummins, Menlo Park, 1987.

54. J. F. MacFadden. Biochemical Tests for the Identification of Medical Bacteria. Williams & Williams, Baltimore, 1980.

55. G. M. Cooper. The Cell: A Molecular Approach. ASM Press, Washington, D.C., 1997.

56. L. Stryer. Biochemistry, 3rd ed. W. H. Freeman, San Francisco, 1990.

57. A. D. Russell. The Destruction of Bacterial Spores. Academic Press, New York, 1982.

58. C. W. Emmons, C. H. Binford, J. P. Utz, & K. J. Kwon-Chung. Medical Mycology, 3rd ed. Lea & Febiger, Philadelphia, 1977.

59. G. S. Moore, & D. M. Jaciow. Mycology for the Clinical Laboratory, Prentice-Hall, Reston, VA, 1979.

60. Y. Al-Doory. Laboratory Medical Mychology, Lea & Febiger, Philadelphia, 1980.

61. J. W. Wilson & O. A. Plunhalt. The Fungous Diseases of Man. University of California Press, Berkeley, 1965.

62. S. G. Dick. Immunological Aspects of Infectious Diseases, University Park Press, Baltimore, 1979.

63. L. S. Kucera & Q. N. Myrvik. Fundamentals of Medical Virulogy, 2nd ed. Lea & Febiger, Philadelphia, 1985.

64. R. C. Weast, Ed. CRC Handbook of Chemistry and Physics, 4th ed. CRC Press, Boca Raton, FL, 1984.

65. S. Budavari. The Merck Index, 12th ed. Merck & Co., Whitehouse Station, NJ, 1996.

66. The United States Pharmacopeia, 23rd ed. The National Formulary (18th ed.). U. S. Pharmacopeial Convention, Inc., Rockville, MD, 1995.

67. W. Gottardi. Iodine and iodine compounds, Disinfection, Sterilization, and Preservation, 4th ed. S. S. Bloch, Ed. Lea & Febiger, Malvern, PA, 152–166, 1991.

68. W. G. Characklis & K. C. Marshall. Biofilms, John Wiley & Sons, New York, 1990.

69. S. F. Bloomfield. Chlorhexidine and iodine formulations, Handbook of Disinfectants and Antiseptics, J. M. Ascenzi, Ed. Marcel Dekker, Inc., New York, 133–158, 1996.

70. G. W. Denton. Chlorhexidine, Disinfection, Sterilization, and Preservation, S. S. Bloch, Ed. Lea & Febiger, Malvern, PA, 274–289, 1991.

71. N. S. Ranganathan.Chlorhexidine, Handbook of Disinfectants and Antiseptics, J. M. Ascenzi, Ed. Marcel Dekker, Inc., New York, 235–264, 1996.

72. M. K. Bruch. Chloroxylenol: An old-new antimicrobial, Handbook of Disinfectants and Antiseptics, J. M. Ascenzi, Ed. Marcel Dekker, Inc., New York, 265–294, 1996.

73. E. L. Larson & H. E. Morton. Alcohols, Disinfection, Sterilization, and Preservation, S. S. Bloch, Ed. Lea & Febiger, Malvern, PA, 191–203, 1991.

74. M. L. Rolter. Alcohols for antisepsis of hands and skin, Handbook of Disinfectants and Antiseptics, J. M. Ascenzi, Ed. Marcel Dekker, Inc., New York, 177–233, 1996.

75. N. K. Denzin & Y. S. Lincoln. Handbook of qualitative research. Sage, Thousand Oaks, CA, 1994.

76. D. C. Montgomery. Design and analysis of experiments, 4th ed. John Wiley & Sons, New York, 1997.

77. W. J. Dixon & F. J. Massey. Introduction to Statistical Analysis, 4th ed. McGraw-Hill, New York, 1983.

78. D. S. Paulson. Statistical Evaluations: How They Can Aid in Developing Successful Cosmetics. Soaps/Cosmetics/Chemical Specialties, Oct. 1985.

79. R. E. Kirk. Experimental Designs, 3rd ed. Brooks/Cole, Pacific Grove, CA, 1995.

80. G. W. Snedecor & W. C. Cochran. Statistical Methods, 7th ed. Iowa State Press, Ames, IO, 1980.

81. P. F. Velleman & D. C. Hoaglin. Applications, basics, and computing of exploratory data analysis. Duxbury Press, Boston, 1981.

82. R. H. Green. Sampling Design and Statistical Methods for Experimental Biologists. John Wiley & Sons, New York, 1979.

83. W. W. Daniel. Applied Nonparametric Statistics. Houghton-Mifflin Co., Boston, 1978.

84. D. S. Paulson. "Marketing Opportunities for Topical Anti-Infective Products," Soaps/Cosmetics/Chemical Specialties, Oct., 31–34, 1994.

85. J. D. Gibbons. Nonparametric Methods for Quantitative Analysis, Holt, Rinehart and Winston, New York, 1976.

86. W. J. Conover. Practical Nonparametric Statistics, 2nd ed. John Wiley & Sons, New York, 1980.

87. D. Polkinghorne. Methodology for the Human Sciences. State University of New York Press, Albany, 1983.

88. W. R. Borg & M. D. Gall. Education Research, 5th ed. Longman, White Plains, New York, 1989.

89. D. S. Paulson. Designing a Handwash Efficacy Program, Pharmaceutical & Cosmetic Quality, April, 17–20, 1996.

90. Annual Book of Standards, Section II, Vol. 11.05, ASTM West Conshohochen, PA, 1998.

91. D. S. Paulson. The Role of Quality Assurance in Effective Topical Antimicrobial Development, Pharmaceutical & Cosmetic Quality, Jan/Feb, 42–44, 1997.
92. D. S. Paulson. Evaluation of Topical Antimicrobial Products. In: Handbook of Disinfectants and Antiseptics, J. M. Ascenzi, ed., Marcel Dekker, Inc., New York, 17–42, 1996.
93. D. G. Kleinbaum & L. L. Kupper. Applied Regression Analysis and Other Multivariable Methods, Duxbury Press, North Scituate, MA, 1978.
94. C. H. Hicks. Fundamental Concepts in the Design of Experiments, 4th ed. Sanders, New York, 1993.

Index

Printed and bound by CPI Group (UK) Ltd, Croydon, CR0 4YY

23/10/2024

01778242-0006